Collins

Student Support Material AQA

540 CHA

A-level Y

Ch

Paper 1 Ino physical ch

Authors: Colin Chambers, Ge ... Maczek, David Nicholls, Rob Symonds, Stephen Whittleton

William Collins' dream of knowledge for all began with the publication of his first book in 1819.

A self-educated mill worker, he not only enriched millions of lives, but also founded a flourishing publishing house. Today, staying true to this spirit, Collins books are packed with inspiration, innovation and practical expertise. They place you at the centre of a world of possibility and give you exactly what you need to explore it.

Collins. Freedom to teach

HarperCollins Publishers
The News Building
1 London Bridge Street
London SE1 9GF

> **Browse the complete Collins catalogue at**
> **www.collins.co.uk**

10 9 8 7 6 5 4 3 2 1

© HarperCollins*Publishers* 2016

ISBN 978-0-00-818950-1

Collins® is a registered trademark of HarperCollins*Publishers* Limited

www.collins.co.uk

A catalogue record for this book is available from the British Library

Thanks to John Bentham and Graham Curtis for their contributions to the previous editions.

Commissioned by Gillian Lindsey
Edited by Alexander Rutherford
Project managed by Maheswari PonSaravanan at Jouve
Development by Tim Jackson
Copyedited and proof read by Janette Schubert
Typeset by Jouve India Private Limited
Original design by Newgen Imaging
Cover design by Angela English
Printed by CPI Group (UK) Ltd, Croydon, CR0 4YY
Cover image © Shutterstock/isaravut

Contents

Introduction

To the student

Book 3 covers Physical Chemistry in sections 3.1.8, 3.1.10, 3.1.11 and 3.1.12 and Inorganic Chemistry in sections 3.2.4, 3.2.5 and 3.2.6.

Questions in each of the A-level papers will assess not only your knowledge and understanding of particular sections of the specification but also your ability to draw together this knowledge and understanding in order to apply it in new and unfamiliar contexts.

A-level Paper 1 can include questions on all of Physical Chemistry, except Kinetics (section 3.1.5) and Rate equations (section 3.1.9), and can also include questions on all of Inorganic Chemistry (as listed in section 3.2).

Questions will also assess your practical experience, including knowledge and understanding of the required practical exercises listed in the Specification.

The Physical Chemistry topics from sections 3.1.1 to 3.1.4, 3.1.6 and 3.1.7 together with the Inorganic Chemistry topics in sections 3.2.1 to 3.2.3 are covered in Book 1. These topics are assessed in AS Paper 1 but also form part of the end-of-course A-level assessment and may also be examined in A-level Paper 1 and A-level Paper 3.

3.1 Physical chemistry

3.1.8 Thermodynamics

3.1.8.1 Born–Haber cycles

Enthalpy change, ΔH

The enthalpy change ΔH was introduced in *Collins Student Support Materials: AS/A-Level year 1 – Inorganic and Relevant Physical Chemistry*, section 3.1.4.

> **Definition**
> The **enthalpy change** is the amount of heat taken in or given out at constant pressure during any physical or chemical change.

Essential Notes

A superscript 'plimsoll' sign, $^{\ominus}$, always follows symbols for quantities measured under standard conditions.

Standard enthalpy change, $\Delta H^{\ominus}$

The size of any enthalpy change depends on the *pressure* and the *temperature* as well as on the *amount of substance* used. Chemists use agreed **standard conditions** in order to make useful comparisons between different measurements and between different sets of data.

The **standard enthalpy change**, $\Delta H^{\ominus}$, is defined as follows:

> **Definition**
> The **standard enthalpy change** for a reaction is the change in enthalpy when reactants in their standard states form products that are also in their standard states.

A **standard state** is defined as follows:

> **Definition**
> The **standard state** of a substance at a given temperature is its pure, most stable, form at 100 kPa.

There remains the need to decide both an *amount of substance* and a *temperature* in order fully to specify any enthalpy change, since enthalpy changes can be quoted for any chosen amount of substance, or for any chosen temperature.

Almost always, the **standard amount** used by chemists is the mole, as introduced in *Collins Student Support Materials: AS/A-Level year 1 – Inorganic and Relevant Physical Chemistry*, section 3.1.2.2. That this standard amount is being used will be perfectly obvious from the units shown, which will almost invariably be quoted as 'per mole' using the symbol **mol^{-1}**.

Notes

The most common reference temperature specified is 298 K.

Most frequently, a **reference temperature** of 298 K is used (see *Collins Student Support Materials: AS/A-Level year 1 – Inorganic and Relevant Physical Chemistry*, section 3.1.4.1). Once this has been specified, ΔH becomes the **standard enthalpy change at 298 K**, shown as $\Delta H^{\ominus}$ (298 K).

However, it is perfectly in order to quote standard quantities at temperatures different from 298 K, if this is convenient, so $\Delta H^{\ominus}$ (1000 K) refers to a standard molar enthalpy change at a temperature of 1000 K. If no temperature is indicated, then it is assumed that the reference temperature is 298 K. Thus, $\Delta H^{\ominus}$ on its own is exactly the same as $\Delta H^{\ominus}$ (298 K).

Some standard enthalpy changes

Some commonly-used standard enthalpy changes are defined below. The standard reference temperature is taken as 298 K in each case.

Standard enthalpy of formation, $\Delta_f H^{\ominus}$

This term was defined in *Collins Student Support Materials:*
AS/A-Level year 1 – Inorganic and Relevant Physical Chemistry, section 3.1.4.1:

> **Definition**
>
> The **standard enthalpy of formation** is the enthalpy change under standard conditions when one mole of a compound is formed from its elements with all reactants and products in their standard states.

Consequently, the standard enthalpy of formation for an element must always be *zero*. Values of standard enthalpy of formation can be found in data books.

Ionisation enthalpy (energy), $\Delta_i H^{\ominus}$

In *Collins Student Support Materials: AS/A-Level year 1 – Inorganic and Relevant Physical Chemistry*, section 3.1.1.3, the *first ionisation energy* of element X was defined as the enthalpy change for the process $X(g) \rightarrow X^+(g) + e^-$. More generally, the ionisation enthalpy is defined as follows:

> **Definition**
>
> The **ionisation enthalpy** is the standard molar enthalpy change for the removal of an electron from a species in the gas phase to form a positive ion and an electron, both also in the gas phase.

This definition embraces all gas phase ionisation processes, not just the **first ionisation enthalpy**, by extending its scope to *any species* (not just to elements), by specifying the *molar quantities* involved, and by referring (correctly, but unimportantly) to electrons in the *gas phase*.

For example, the *ionisation enthalpy* of sodium refers to the process:

$$Na(g) \rightarrow Na^+(g) + e^-(g) \qquad \Delta_i H^{\ominus} = +496 \text{ kJ mol}^{-1}$$

For a neutral species, such as sodium, which loses just a single electron, this is called the first ionisation enthalpy.

The cation formed, $Na^+(g)$, may itself be further ionised, according to the process:

$$Na^+(g) \rightarrow Na^{2+}(g) + e^-(g) \qquad \Delta_i H^{\ominus} = +4562 \text{ kJ mol}^{-1}$$

This is called the **second ionisation enthalpy** of sodium. Some typical values of first and second ionisation enthalpies are shown in Table 1.

Notes

The IUPAC-recommended symbol for enthalpy of combustion (defined as the enthalpy change on combustion) is $\Delta_c H$. In older textbooks you may see this written as ΔH_c. This symbol, with 'c' for combustion shown after 'H' for enthalpy, represents the same enthalpy change but is no longer the recommended format. In this book we will represent all enthalpy changes in the latest format as recommended by IUPAC.

Notes

The positive ion is called a cation; its negative counterpart is an anion.

Notes

The state symbol (g) for the electron can be assumed and is often omitted.

Notes

Enthalpy change is measured at **constant pressure**. Ionisation involves the formation of 2 mol of gaseous products from 1 mol of gaseous reactants, so there is a subtle difference between *ionisation enthalpy* and *ionisation energy*, but this can be ignored here. A similar small difference exists in the case of electron affinity; this too can be ignored.

$\Delta_iH^\ominus$ / kJ mol^{-1}		
	First	**Second**
H(g)	1310	–
He(g)	2370	5250
Mg(g)	736	1450
Na(g)	494	4560

Table 1
Ionisation enthalpy at 298 K

Note that the second ionisation enthalpy of an atom is *always* larger than the first, because the removal of the second electron from an already positively-charged species requires more energy than the removal of the first from a neutral species.

Enthalpy of atomisation, $\Delta_{at}H^\ominus$

> **Definition**
>
> The **enthalpy of atomisation** is the standard enthalpy change that accompanies the formation of one mole of gaseous atoms from the element in its standard state.

For an atomic solid, such as an element, the standard enthalpy of atomisation is simply the standard enthalpy of sublimation of the solid. For example:

$$Na(s) \rightarrow Na(g) \qquad\qquad \Delta_{sub}H^\ominus = +107 \text{ kJ mol}^{-1}$$

In such a case, the enthalpy of atomisation is identical to the enthalpy of sublimation:

$$\Delta_{at}H^\ominus = \Delta_{sub}H^\ominus$$

Enthalpies of atomisation (**sublimation**) for a selection of substances are shown in Table 2. Sublimation always requires an input of energy (*endothermic* process), so these enthalpies are all positive.

$\Delta_{at}H^\ominus$ / kJ mol^{-1}	
C(graphite)	715
Na(s)	109
K(s)	90
Mg(s)	150

Table 2
Enthalpy of atomisation

In the case of bond fission, a diatomic molecule will produce two moles of atoms, so the enthalpy of atomisation is half the bond dissociation enthalpy (see Table 3). Thus, for chlorine:

$$\tfrac{1}{2}Cl_2(g) \rightarrow Cl(g) \qquad\qquad \Delta_{at}H^\ominus = \tfrac{1}{2}(+242) = +121 \text{ kJ mol}^{-1}$$

so $\quad \Delta_{at}H^\ominus = \tfrac{1}{2}\Delta_{diss}H^\ominus$

Bond dissociation enthalpy, $\Delta_{diss}H^\ominus$

> **Definition**
>
> The **bond dissociation enthalpy** is the standard molar enthalpy change that accompanies the breaking of a covalent bond in a gaseous molecule to form two gaseous free radicals.

Essential Notes

Homolytic bond fission occurs when a covalent bond breaks and the shared electron pair is split equally between the resulting species (*free radicals*). In **heterolytic** bond fission, both the shared electrons end up on one of the resulting species (*ions*).

In order to indicate that each species formed by bond fission has one unpaired electron, it is considered correct, particularly in processes involving bond fission in a mass spectrometer or in radical chain reactions, to write a *dot* alongside the odd-electron species to indicate the unpaired electron. This species is called a **free radical**, e.g. the methyl radical $\bullet CH_3$ or the chlorine atom $\bullet Cl$.

> **Definition**
>
> A **free radical** is a species that results from the homolytic fission of a covalent bond. It contains an unpaired electron, since homolytic fission results in the splitting of the electron pair in a covalent bond, one electron going to each partner.

In thermodynamic equations involving bond fission, it is quite common to omit *dots* representing unpaired electrons, since their presence is obvious from the equation as written, e.g.

$$Cl_2(g) \rightarrow 2Cl(g) \qquad \Delta_{diss}H^{\ominus} = +242 \text{ kJ mol}^{-1}$$

or

$$CH_4(g) \rightarrow CH_3(g) + H(g) \qquad \Delta_{diss}H^{\ominus} = +435 \text{ kJ mol}^{-1}$$

Some bond dissociation enthalpies for homolytic fission in a selection of covalent compounds are shown in Table 3.

$\Delta_{diss}H^{\ominus}$ / kJ mol^{-1}			
H–H	436	N≡N	945
H–F	565	O=O	496
H–Cl	431	F–F	158
H–Br	366	Cl–Cl	242
H–I	299	Br–Br	194

Table 3
Bond dissociation enthalpy

Electron affinity, $\Delta_{ea}H^{\ominus}$

Definition

Electron affinity is the standard enthalpy change when an electron is added to an isolated atom in the gas phase.

Electron affinity refers to a process of the type:

$$Cl(g) + e^-(g) \rightarrow Cl^-(g) \qquad \Delta_{ea}H^{\ominus} = -364 \text{ kJ mol}^{-1}$$

A chlorine atom in the gas phase has a strong **affinity** for an electron, so that the capture of an electron to form a gaseous chloride ion causes energy to be given out to the surroundings – an *exothermic* process.

Some typical values of electron affinity are shown in Table 4.

The enthalpy of formation of $O^{2-}(g)$ from $O^-(g)$ is +844 kJ mol^{-1}. The process is strongly endothermic, since it takes a lot of energy to force a second electron onto the already negative O^- ion.

$\Delta_{ea}H^{\ominus}$ / kJ mol^{-1}	
H(g)	–72
F(g)	–348
Cl(g)	–364
Br(g)	–342
O(g)	–142
O^-(g)	+844

Table 4
Electron affinity at 298 K

Lattice enthalpy, $\Delta_L H^{\ominus}$

Definition

The *enthalpy of lattice dissociation* is the standard enthalpy change that accompanies the separation of one mole of a solid ionic lattice into its gaseous ions.

The *enthalpy of lattice formation* is the converse of this, i.e. the standard enthalpy change that accompanies the formation of one mole of a solid ionic lattice from its gaseous ions.

Notes

The change $O^-(g) + e^-(g) \rightarrow O^{2-}(g)$ is known as the second electron affinity of oxygen. The term **electron affinity** on its own always refers to the (first) electron affinity, as defined above.

For example:

$$NaCl(s) \rightarrow Na^+(g) + Cl^-(g) \qquad \Delta_L H^{\ominus} = +771 \text{ kJ mol}^{-1}$$

If *lattice dissociation* is used as a defining equation, as above, all lattice enthalpies are *positive*, since all ionic crystals are energetically more favoured than their separated gaseous ions. Consequently, it requires an *input* of energy to disrupt the crystal lattice and form separated gaseous ions.

Notes

Bond making and lattice formation are exothermic.

Bond breaking and lattice dissociation are endothermic.

$\Delta_L H^\ominus$ / kJ mol^{-1}	
NaF(s)	902
NaCl(s)	771
KCl(s)	701
MgO(s)	3889
MgS(s)	3238

Table 5
Lattice dissociation enthalpy values at 298 K

$\Delta_{hyd} H^\ominus$ / kJ mol^{-1}			
Li$^+$	−519	F$^-$	−506
Na$^+$	−406	Cl$^-$	−364
K$^+$	−322	Br$^-$	−335

Table 6
Hydration enthalpy values for individual ions

If the reverse defining equation is used:

$$Na^+(g) + Cl^-(g) \rightarrow NaCl(s) \qquad \Delta_L H^\ominus = -771 \text{ kJ mol}^{-1}$$

then the process of interest is *lattice formation*, and the resulting *enthalpy of lattice formation* is always *negative*. Be aware that such conflicting definitions exist and be able to distinguish which definition is in use from the sign of the resulting lattice enthalpy or the direction of the arrow in the defining equation.

Some typical lattice dissociation enthalpy values are shown in Table 5.

Enthalpy of hydration, $\Delta_{hyd} H^\ominus$

Definition

The **enthalpy of hydration** is the standard enthalpy change for the process:

$$X^\pm(g) \xrightarrow{\text{water}} X^\pm(aq) \qquad \Delta H^\ominus = \Delta_{hyd} H^\ominus$$

where the single symbol, $X^\pm(g)$ is used to indicate both a gaseous cation, $X^+(g)$, and a gaseous anion, $X^-(g)$, with $X^\pm(aq)$ denoting the appropriate hydrated species in solution.

Hydration enthalpy values for a selection of ions are shown in Table 6.

Enthalpy of solution, $\Delta_{sol} H^\ominus$

Definition

The **enthalpy of solution** is the standard enthalpy change that occurs when one mole of an ionic solid dissolves in enough water to ensure that the dissolved ions are well separated and do not interact with one another.

For example:

$$NaCl(s) \xrightarrow{\text{water}} Na^+(aq) + Cl^-(aq) \qquad \Delta_{sol} H^\ominus = +2 \text{ kJ mol}^{-1}$$

The Born–Haber cycle

The enthalpy of formation of a solid ionic compound can be broken up into a number of steps, arranged in a cycle which is called the **Born–Haber cycle**. Enthalpy cycles were introduced earlier (*Collins Student Support Materials: AS/A-Level year 1 – Inorganic and Relevant Physical Chemistry*, section 3.1.4.3); they make use of Hess's Law to establish that the sum of the values of all the changes, travelling in either direction round the cycle, must equal zero. This provides a convenient method to find the enthalpy of an unknown step in the cycle if the enthalpy changes for all other steps are known.

Notes

Enthalpy cycles: A cycle consists of a series of different enthalpy changes which, starting at any point in the cycle, then end up at the same starting point. The sum of all the enthalpy changes round a cycle is zero, $\sum_{cycle} \Delta H = 0$.

The Born–Haber cycle includes a step involving the enthalpy of lattice formation, $\Delta_L H^\ominus$, the value of which cannot be determined experimentally, so it is possible to deduce this lattice enthalpy using the Born–Haber cycle and other thermodynamic data. The procedure used is shown in Fig 1.

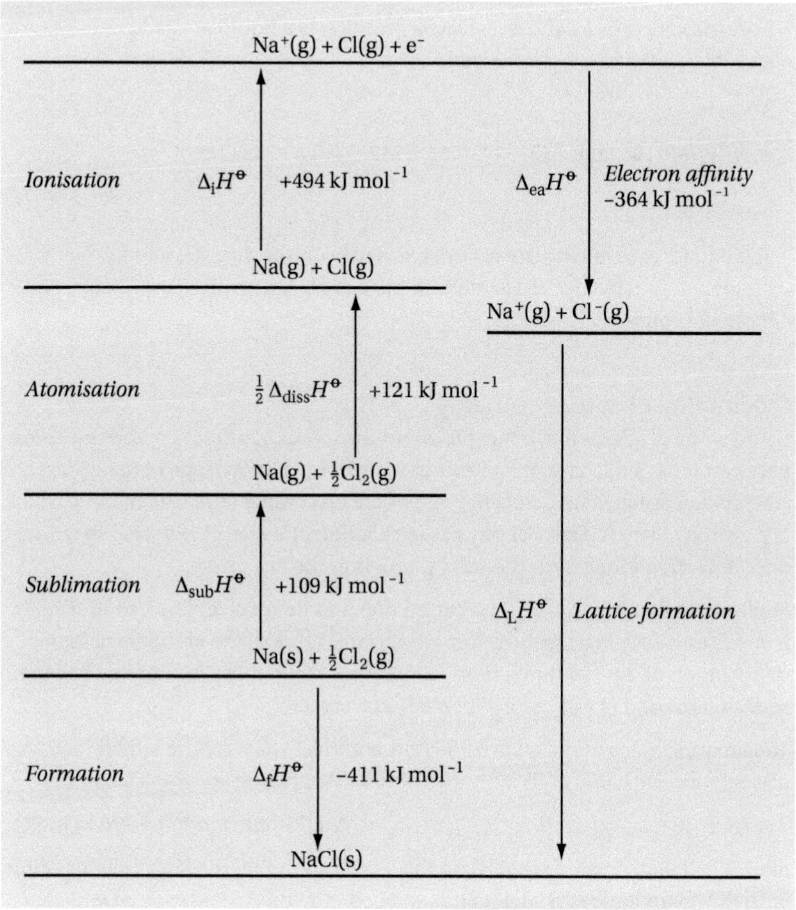

Fig 1
A Born–Haber cycle for the determination of the lattice enthalpy of sodium chloride. The sum of the enthalpy changes round the cycle is zero, so that the magnitude of any unknown enthalpy in the cycle can be found if all the other enthalpies are known

Notes

The Born–Haber cycle: starting from the elements, proceed to form the compound by just one step (directly) or by several steps (indirectly).

Notes

The term *sublimation* refers to the process in which a solid changes directly into a gas (e.g. $I_2(s) \rightarrow I_2(g)$) without going through an intermediate liquid phase. Note that this term is not included in the AQA A-level specification, but it does provide a useful form of shorthand.

Applying Hess's Law:

$$\Delta H^\ominus \text{ (single step)} = \Delta H^\ominus \text{ (five steps)}$$

So $\quad \Delta_f H^\ominus \qquad = + \Delta_{sub}H^\ominus + \tfrac{1}{2}\Delta_{diss}H^\ominus + \Delta_i H^\ominus + \Delta_{ea}H^\ominus + \Delta_L H^\ominus$

and $\quad \Delta_L H^\ominus \qquad = - \Delta_{sub}H^\ominus - \tfrac{1}{2}\Delta_{diss}H^\ominus - \Delta_i H^\ominus - \Delta_{ea}H^\ominus + \Delta_f H^\ominus$

Thus $\quad \Delta_L H^\ominus(\text{NaCl}) = -109 - 121 - 494 - (-364) + (-411)$

$$= -771 \text{ kJ mol}^{-1}$$

Notes

The value of the standard enthalpy of formation of NaCl(s) used in this calculation is found from data tables and is -411 kJ mol^{-1}.

Example

Calculate the lattice enthalpy of sodium bromide. Use data taken from tables in the text. The **enthalpy of vaporisation** of bromine is $+30 \text{ kJ mol}^{-1}$ and the standard enthalpy of formation of solid sodium bromide is -360 kJ mol^{-1}.

Method

Construct a cycle, as above. Because bromine is a liquid at 298 K, an extra equation/step is needed.

Answer

$\Delta_L H^{\ominus}(\text{NaBr}) = -109 - 15 - 97 - 494 + 342 - 360 = -733 \text{ kJ mol}^{-1}$

Comment

The equation involving the vaporisation of bromine, like the one for the dissociation to form a single atom in the gas phase, involves only half a mole of bromine.

Calculation of lattice enthalpy

Using some fundamental concepts about the forces of attraction and repulsion between ions of like and unlike charges, and with a knowledge of the geometry (the distances between adjacent ions) of the crystalline lattice of an ionic salt, it is possible to use the laws of physics to calculate the energy required to convert one mole of salt ions from the solid phase into the gas phase.

Such a calculation is based on what is known as the **perfect ionic model** and it yields an answer to the question: what is the value of the enthalpy of lattice dissociation of, say, sodium chloride, assuming that only ionic forces hold the sodium ions and chloride ions together in a crystal?

The answer to this question provides a theoretical value for the lattice dissociation enthalpy $\Delta_L H$:

$$\text{NaCl(s)} \rightarrow \text{Na}^+\text{(g)} + \text{Cl}^-\text{(g)} \qquad \Delta_L H^{\ominus}(\text{ionic model}) = +76 \text{ kJ mol}^{-1}$$

Such a calculated value is based on the assumption that the lattice dissociation enthalpy of an ionic crystal depends solely on the total separation of ions that experience only purely ionic (electrostatic) forces. This ionic model makes it possible to allow a comparison between:

- an experimentally-based value of the lattice dissociation enthalpy (found by using a Born–Haber cycle) and

- a theoretically-based value, calculated by assuming that ionic (electrostatic) forces are solely responsible for the stability of ionic crystalline solids.

With these two methods (experiment and calculation) available, the first thing to do is to see if the results agree and, if so, how well and for which compounds. Given the approximations that are needed in making valid calculations, agreement needs only to be close, not exact.

The most relevant first point of comparison between experiment and calculation is for the alkali metal halides. Table 7 shows values of experimental *Born–Haber* (*BH*) lattice dissociation enthalpies alongside the corresponding calculated *ionic model* (*IM*) values.

Notes

The questions raised here, as well as their answers, provide a graphic illustration of *How Science Works*. Data from one source (experiment + Born–Haber) are compared with potentially equally valid data from another source (perfect ionic model theory + calculation). Agreement confers a degree of validity on the assumptions made in the theoretical model; disagreement provides a motive and a pressing need to ask further questions and to seek alternative explanations.

$\Delta_L H$ / kJ mol^{-1}							
F		Cl		Br		I	
BH	IM	BH	IM	BH	IM	BH	IM
Li	1022	1004	846	833	800	787	744	728
Na	902	891	771	766	753	752	684	686
K	801	795	701	690	670	665	629	632

Table 7
Experimental (**BH**) and calculated (**IM**) lattice dissociation enthalpies for some alkali metal halides

There is very satisfactory agreement between the Born–Haber (experimental) values and the perfect ionic model (theoretical) values in this table. The average discrepancy amounts to just over 8 kJ mol^{-1}, which is roughly of the order of 1%. It is therefore reasonable to conclude that these simple alkali halide ionic crystals behave just as would be expected for systems consisting of oppositely-charged discrete ions interacting exclusively through electrostatic attractions and repulsions. The ionic model works almost perfectly.

However, such good agreement does not persist for other presumed ionic crystals. Table 8 contains data for the silver halides.

$\Delta_L H$ / kJ mol^{-1}							
	F		Cl		Br		I	
	BH	IM	BH	IM	BH	IM	BH	IM
Ag	955	870	905	770	890	758	876	736

Table 8
Experimental (**BH**) and calculated (**IM**) lattice dissociation enthalpies for the silver halides

Here, there is far less agreement. The average discrepancy amounts to just over 120 kJ mol^{-1}, or some 13–14%. Some other factor must be having an influence.

A simple explanation of these discrepancies is that the ionic model takes no account of such factors as a degree of covalent bonding between the constituent ions. Such additional bonding would require an additional input of energy in order to separate the constituents of the solid into free separate gaseous ions. This requirement is consistent with the considerably greater value of the experimental lattice dissociation enthalpy (*BH*) as compared with the purely ionic calculation (*IM*).

The increased lattice enthalpy of the silver halides compared with the ionic model may account for their lack of solubility in water. It might be expected that silver chloride, were it a simple ionic compound like sodium chloride, would be soluble in water. However, the additional heat energy required to separate the ions before they are hydrated makes the overall free-energy change for this process positive (see this book, section 3.1.8.2), so that dissolving is no longer a 'feasible' process.

Calculating enthalpies of solution
The following example shows a calculation that uses enthalpies of hydration and lattice enthalpy.

Notes
Experimental values of the lattice dissociation enthalpy seem almost always to be greater than those calculated using the *perfect ionic model*. This discrepancy can be accounted for if it is assumed that some presumed ionic compounds possess a degree of covalent bonding (*covalent character*) which adds stability to the solid state and thus requires additional energy in converting the solid into independent gaseous ions.

Example

Calculate the enthalpy of solution of sodium chloride, using data taken from tables in the text.

Answer

Method 1

Draw an enthalpy cycle as shown below.

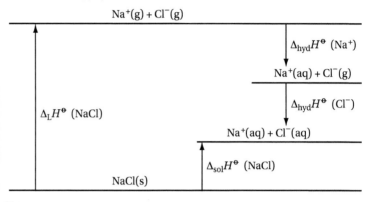

Notes

The arrow in the diagram leads from solid NaCl to separated gaseous ions. Clearly, $\Delta_L H^{\ominus}$ (NaCl) in this case is the *enthalpy of lattice dissociation.*

Hence

$$\Delta_{sol}H^{\ominus} \ (NaCl) = \Delta_L H^{\ominus} \ (NaCl) + \Delta_{hyd}H^{\ominus} \ (Na^+) + \Delta_{hyd}H^{\ominus} \ (Cl^-)$$

Therefore

$$\Delta_{sol}H^{\ominus} \ (NaCl) = \quad +771 \quad + \quad (-405) \quad + \quad (-364) \quad = +2 \text{ kJ mol}^{-1}$$

Method 2

Write equations, including enthalpy changes (reversed where appropriate) for lattice enthalpy and hydration enthalpies in such a way that simple addition yields the equation for enthalpy of solution. Addition of the enthalpy changes gives the overall enthalpy of reaction.

$NaCl(s) \rightarrow Na^+(g) + Cl^-(g)$	$\Delta_L H^{\ominus} \quad = +771 \text{ kJ mol}^{-1}$
$Na^+(g) \xrightarrow{\text{water}} Na^+(aq)$	$\Delta_{hyd}H^{\ominus} = -405 \text{ kJ mol}^{-1}$
$Cl^-(g) \xrightarrow{\text{water}} Cl^-(aq)$	$\Delta_{hyd}H^{\ominus} = -364 \text{ kJ mol}^{-1}$
$NaCl(s) \xrightarrow{\text{water}} Na^+(aq) + Cl^-(aq)$	$\Delta_{sol}H^{\ominus} = +2 \text{ kJ mol}^{-1}$

Comment

Either method may be used with confidence. The one that is chosen is entirely a matter of personal preference, although *Method 2* is slightly more transparent and marginally less prone to mistakes.

Note also that the enthalpy of solution is *very* small, and is calculated here as the difference between two very large quantities. Such calculations are often very unreliable and are subject to round-off errors, particularly where other enthalpies are quoted to only three significant figures. Their difference, therefore, is good only to one significant figure.

3.1.8.2 Gibbs free-energy change, ΔG, and entropy change, ΔS

Spontaneous change

Definition

A spontaneous change is one that has a natural tendency to occur without being driven by any external influences.

Spontaneous changes are familiar from everyday life. The air compressed in a bicycle tyre escapes spontaneously if the valve is removed, and a lot of physical effort is needed to pump it up again. Hot soup cools spontaneously, and heat energy from a gas ring is needed to warm it up again. Ice cream melts spontaneously on a hot day, and needs the electrical energy put into a refrigerator to freeze it again. Iron rusts spontaneously in damp air, and it takes all the vast energy of a blast furnace to get iron back from an oxide ore. A rechargeable dry cell can light the bulb of a torch spontaneously, but has to be left to draw energy from the mains when being recharged.

Left to themselves, therefore, some things simply happen. Others do not happen unless energy is expended. A **spontaneous change** is one that can occur in one particular direction but not in reverse (unless conditions such as temperature are changed).

Notes

If a process in one direction is spontaneous, the reverse process will require an input of external energy.

The enthalpy factor

Many spontaneous chemical reactions appear to be driven by a favourable change in enthalpy. In all the changes listed above, it would be easy to reach the simple (but mistaken) conclusion that spontaneous change occurs if heat is given out as the change is taking place. But this is not the whole story.

Exothermic reactions are often spontaneous. The natural direction of spontaneous change is from higher to lower enthalpy, with a release (exothermic) of the difference in energy between the two. For example, the reaction:

$$H_2(g) + \tfrac{1}{2}O_2(g) \rightarrow H_2O(g) \qquad \Delta H^{\ominus} = -242 \text{ kJ mol}^{-1}$$

liberates vast quantities of heat energy. The reaction is exothermic, the enthalpy change is negative ($\Delta H < 0$), and this favours spontaneous reaction.

It would be tempting to conclude that spontaneous reactions must involve a release of heat energy. However, some reactions are spontaneous even though they are **endothermic**. For example, the reaction:

$$KHCO_3(s) + HCl(aq) \rightarrow$$
$$KCl(aq) + H_2O(l) + CO_2(g) \qquad \Delta H^{\ominus} = +25 \text{ kJ mol}^{-1}$$

is endothermic and the temperature of the reaction mixture drops when the reactants are mixed. Yet it proceeds spontaneously. Clearly, there must be some additional factor that causes reactions to occur, over and above the simple release of heat energy.

Notes

Most spontaneous reactions are exothermic, but only some endothermic reactions are spontaneous.

The entropy factor

The additional factor that helps to drive spontaneous change in a given direction is called the **entropy**, which is given the symbol **S**.

Before proceeding, in the next sections, to relate entropy to everyday experience, it is useful to make a few statements about entropy and its nature,

Essential Notes

Entropy (S) has the units $J\,K^{-1}\,mol^{-1}$.
Note that J rather than kJ is used.

just to get an insight into this new quantity that turns out to be one of the driving forces in spontaneous change.

(a) Entropy and disorder

It proves very helpful to *think of* entropy in terms of *disorder*. An increase in entropy can be visualised as an increase in disorder. Processes leading to increased chaos are rather more likely than those leading to order. Heating leads to an increase in disorder, which fits with the increase in entropy as a substance changes from solid, to liquid, to gas.

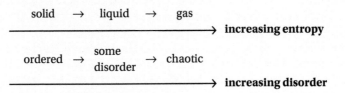

(b) Variation of entropy with temperature

A graph showing typical changes of entropy as temperature is increased is shown below.

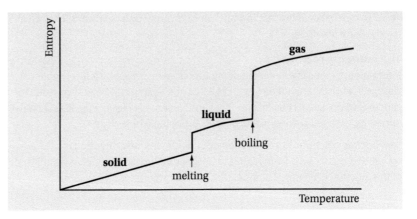

It is clear that:

- entropy increases slowly with temperature in solids, liquids and gases
- a phase change (*change of physical state*) causes a sudden change in entropy
- boiling causes a larger increase in entropy than does melting.

(c) Magnitude of entropy in different substances

In Table 9, some **standard entropy** values of different substances are listed. Two broad trends can be noted:

- simple molecules generally seem to have lower entropies than more complicated molecules, for example:

$$S^{\ominus}(CO) < S^{\ominus}(CO_2); \quad S^{\ominus}(He) < S^{\ominus}(O_2); \quad S^{\ominus}(NaCl) < S^{\ominus}(NaHCO_3)$$

- for substances of similar complexity, entropy generally increases on going from solid, to liquid, to gas, for example:

$$S^{\ominus}(H_2O(s)) < S^{\ominus}(H_2O(l)) < S^{\ominus}(H_2O(g)); \quad S^{\ominus}(CH_3OH(l)) < S^{\ominus}(CH_3OH(g))$$

$S^{\ominus}_{298}$/J K^{-1} mol^{-1}			
diamond	2.4	He(g)	126
graphite	5.7	Ar(g)	155
NaCl(s)	72	O_2(g)	205
NaHCO$_3$(s)	102	CO(g)	198
SiO$_2$(s)	42	CO$_2$(g)	214
H$_2$O(l)	70	H$_2$O(g)	189
CH$_3$OH(l)	127	CH$_3$OH(g)	240

Table 9
Standard entropy values at 298 K

(d) Units in which S is measured

In the SI system, the units of entropy are $J\,K^{-1}\,mol^{-1}$. In themselves, these units say little that is new about entropy, but they will prove useful and revealing later on when the separate influences of enthalpy and entropy are combined in determining the direction of spontaneous change.

(e) Absolute entropy, $S^{\ominus}$

Since entropy is so closely linked to disorder, and as it decreases when the temperature is lowered, it is reasonable to suggest that, at 0 K (the absolute zero), all disorder will vanish and all substances will be perfectly ordered and have *zero entropy*. This conclusion turns out to be true for most substances, and especially for perfectly-ordered crystals.

The concept of zero entropy has one important consequence. With enthalpy, it is possible to consider only standard enthalpy *changes*, $\Delta H^{\ominus}$ (*i.e. only differences between enthalpy values*); in the case of entropy, there is a well-defined starting point to the entropy scale (zero entropy at zero Kelvin). As a consequence, it is possible to speak of and to list values for *absolute standard entropies, $S^{\ominus}$*, and from these to calculate *standard entropy changes, $\Delta S^{\ominus}$*. Table 9 lists some **absolute entropy values**.

> **Notes**
>
> Absolute entropy values are based on the premise that all substances have zero entropy at the absolute zero of temperature.

Entropy change in chemical reactions, $\Delta S^{\ominus}$

Standard entropy changes in systems undergoing chemical change are calculated as the difference between the entropy of the products and the entropy of the reactants. Absolute entropy values of products and reactants are obtained from tables of *standard entropies* such as Table 9.

Entropy calculations using standard entropy values use the equation:

$$\Delta S^{\ominus} = \Sigma S^{\ominus}_{products} - \Sigma S^{\ominus}_{reactants}$$

The following examples illustrate how such calculations are performed.

Example

Calculate the entropy change that accompanies the combustion of graphite. Use the entropy data in Table 9.

Method

The reaction in question is:

$$C(s) + O_2(g) \rightarrow CO_2(g)$$

Answer

$$\Delta S^{\ominus} = \Sigma S^{\ominus}_{products} - \Sigma S^{\ominus}_{reactants}$$

$\Sigma S^{\ominus}_{products}$	$= 214.0\ J\,K^{-1}\,mol^{-1}$
$\Sigma S^{\ominus}_{reactants} = 5.7 + 205$	$= 210.7\ J\,K^{-1}\,mol^{-1}$
So $\Delta S^{\ominus} \qquad = 214.0 - 210.7$	$= +3.3\ J\,K^{-1}\,mol^{-1}$

Comment

The entropy change here is quite small. Carbon dioxide and oxygen are both gases and have similar entropies. Graphite is a very low-entropy solid. So even though two moles form one, the number of gas moles does not change and there is not much change in entropy.

Example

Calculate the entropy change that accompanies the decomposition by heating of solid sodium hydrogencarbonate. Use the entropy data given in Table 9, and the additional information that solid sodium carbonate has a standard entropy of 136 J K^{-1} mol^{-1}.

Method

The reaction in question is:

$$2NaHCO_3(s) \rightarrow Na_2CO_3(s) + H_2O(g) + CO_2(g)$$

Answer

$$\Delta S^{\ominus} = \Sigma S^{\ominus}_{products} - \Sigma S^{\ominus}_{reactants}$$

$$\Sigma S^{\ominus}_{products} = 136 + 189 + 214 = 539 \text{ J K}^{-1}\text{ mol}^{-1}$$

$$\Sigma S^{\ominus}_{reactants} = 2 \times 102 \qquad = 204 \text{ J K}^{-1}\text{ mol}^{-1}$$

So $\Delta S^{\ominus}$ $\qquad = 539 - 204 \qquad = +335 \text{ J K}^{-1}\text{ mol}^{-1}$

Essential Notes

The unit mol^{-1} (*'per mole'*) means *'per mole of the defining equation, as written'*.
If the defining equation is doubled, then all the associated thermodynamic quantities are also doubled.
All *molar* thermodynamic quantities are tied, uniquely, to their own defining equations.

Comment 1

The entropy change calculated is *per mole of equation as written*, which here refers to the decomposition of 2 mol of NaHCO$_3$. For one mole of NaHCO$_3$, ΔS is halved, giving 167.5 J K^{-1} mol^{-1}.

Comment 2

The entropy change here is very much larger than that found in the previous example. In the case above, 2 mol of gas are formed from a solid, so the disorder of the system is increased much more.

Gibbs free-energy change, $\Delta G^{\ominus}$, spontaneous change and feasibility

It is clear that reactions in which there is a release of energy (**exothermic processes**) tend to happen quite often. Equally, reactions in which there is an intake of energy (**endothermic processes**) but which lead to an increase in disorder (*increase in entropy*) also happen quite often. The sometimes conflicting demands of enthalpy and entropy are brought together in the relationship:

$$\Delta G^{\ominus} = \Delta H^{\ominus} - T\Delta S^{\ominus}$$

where the standard free-energy change, $\Delta G^{\ominus}$, combines the influence of both enthalpy and entropy changes.

This equation expresses the benefit of an exothermic reaction (ΔH *negative*) with one in which entropy increases (ΔS *positive*); both of these conditions are associated with changes that do happen, leading to the conclusion that:

ΔG must be negative (or zero) for spontaneous change.

For example, the reaction between gaseous hydrogen and oxygen at room temperature is classed as **spontaneous** (or **feasible**) because ΔG for the formation of water from its elements is very negative. So it *can* go. The reaction *will not* go, however, unless a spark or a flame is applied, whereupon an explosion results.

Similarly, hydrogen and chlorine do not react in the dark, but do so with explosive violence in sunlight.

In such cases, an **activation energy** barrier (see *Collins Student Support Materials: AS/A-Level year 1 – Organic and Relevant Physical Chemistry*, section 3.1.5.1) has to be overcome before reaction can occur.

Spontaneous ≡ feasible

In thermodynamics, the words *spontaneous* and *feasible* have exactly the same meaning. In everyday speech, however, the word *spontaneous* (as in *spontaneous applause* or *spontaneous tears*) implies something that is almost *inevitable*, that simply *must* happen. In thermodynamics, spontaneity has to do only with a *tendency* for something to happen, and not at all if it *actually will* happen. The term *'feasible'* is much to be preferred, implying as it does, something that is possible but not inevitable, and it is the term that will be used in what now follows. However, it is worth noting that the term *'spontaneous'* is firmly enshrined in the language of equilibrium and that it frequently provides an acceptable alternative to *feasible*.

Quite often, the speed at which a feasible change occurs can be rather slow (as in the cooling of a cup of coffee), or even infinitely slow (as in the change from diamond to graphite). Nonetheless, both these changes (cold coffee and no diamonds) are truly *feasible* because a decrease in free energy occurs.

Thermodynamics has nothing to say about the speed with which things happen: only whether they *can* happen. Thermodynamics answers the first and fundamental question: *can it go?* (is the reaction *feasible*, is $\Delta G < 0$?). Only if the answer to this first question is *'yes'* is it worth posing the second question: *how fast will it go?* (are the *kinetics* favourable?).

A chemical or physical change is said to be *feasible* if the value of ΔG is *negative* or *zero*. When ΔG is *positive*, the change is said to be *unfeasible* (or *not feasible*).

The way to think of *feasibility*, and specifically of the criterion $\Delta G = 0$, is to understand that $\Delta G = 0$ defines a break-even point where, on balance, there will be similar concentrations of reactants and products present at equilibrium. If $\Delta G < 0$, there will be more products than reactants at equilibrium; if $\Delta G > 0$, reactants will predominate.

Notes

A relatively small change in the value of ΔG can tip equilibrium from being largely favourable to being rather unfavourable. At 300 K, as little as a ± 6 kJ mol^{-1} change in ΔG can tip an equilibrium from 10:1 in favour of reactants to 10:1 in favour of products.

Calculations involving $\Delta G^{\ominus}$

From standard enthalpy and standard entropy data, the standard free-energy change can be calculated using $\Delta G = \Delta H - T\Delta S$, from which the feasibility of a chemical reaction can be determined. This procedure is illustrated in the examples that follow. The data required are given in Table 10.

Table 10
Standard enthalpy and entropy changes for three different chemical reactions

Chemical reaction	$\Delta_{298}H^{\ominus}$ /kJ mol^{-1}	$\Delta_{298}S^{\ominus}$ /J K^{-1} mol^{-1}
$C(s) + O_2(g) \rightarrow CO_2(g)$	−394	+3.3
$2Fe(s) + \frac{3}{2}O_2(g) \rightarrow Fe_2O_3(s)$	−825	−272
$2NaHCO_3(s) \rightarrow Na_2CO_3(s) + H_2O(g) + CO_2(g)$	+130	+335

Notes

Note that the first reaction is always feasible, that low temperatures favour the second, while the third is favoured by high temperatures.

Example

Calculate the standard free-energy change for the combustion of graphite at 298 K. Use the data given in Table 10.

The reaction in question is:

$$C(s) + O_2(g) \rightarrow CO_2(g)$$

Answer

$$\Delta G^{\ominus} = \Delta H^{\ominus} - T\Delta S^{\ominus}$$
$$\Delta_{298}G^{\ominus} = -394 \text{ kJ mol}^{-1} - \frac{298 \times 3.3}{1000} \text{ kJ mol}^{-1}$$
$$= -395 \text{ kJ mol}^{-1}$$

Notes

ΔG is negative, so the reaction is *feasible*. The entropy term is very small so $\Delta G \equiv \Delta H$. Because ΔS is so small, ΔG will not vary much with temperature, though in practice the reaction is extremely slow at 298 K (high activation energy).

Comment

The energy units of ΔH and $T\Delta S$ must be made compatible using division by 1000 to convert the value of ΔS (given in J K^{-1} mol^{-1}) into kJ K^{-1} mol^{-1}. This operation renders the entropy term (3.3×10^{-3} kJ K^{-1} mol^{-1}) vanishingly small, even when multiplied by 298 K.

Example

Calculate the standard free-energy change for the rusting of iron at 298 K. Use the data given in Table 10.

The reaction in question is:

$$2Fe(s) + \frac{3}{2}O_2(g) \rightarrow Fe_2O_3(s)$$

Answer

$$\Delta G^{\ominus} = \Delta H^{\ominus} - T\Delta S^{\ominus}$$
$$\Delta_{298}G^{\ominus} = -825 \text{ kJ mol}^{-1} - \frac{298 \times -272}{1000} \text{ kJ mol}^{-1}$$
$$= -744 \text{ kJ mol}^{-1}$$

Notes

ΔG here is very negative, so the reaction is *feasible*. The entropy change is negative because disorder in the gas is lost forming a solid. ΔG is negative, however, because the exothermic value of ΔH dominates.

Comment

ΔG is hugely negative, so that the rusting of iron is highly feasible at room temperature, in complete accord with everyday experience. The reaction is favoured by low temperatures but does not cease to be feasible until the temperature exceeds 3000 K. It is strongly driven by a very exothermic enthalpy change.

Example

Calculate the standard free-energy change for the decomposition of one mole of sodium hydrogencarbonate at 298 K. Use the data in Table 10.

The reaction in question is:

$$2NaHCO_3(s) \rightarrow Na_2CO_3(s) + H_2O(g) + CO_2(g)$$

Answer

$$\Delta G^{\ominus} = \Delta H^{\ominus} - T\Delta S^{\ominus}$$

$$\Delta G^{\ominus} = 130 \text{ kJ mol}^{-1} - \frac{298 \times 335}{1000} \text{ kJ mol}^{-1} = +30 \text{ kJ mol}^{-1}$$

So, for 1 mol of $NaHCO_3$, $\Delta_{298}G^{\ominus} = \frac{1}{2} \times 30 = +15 \text{ kJ mol}^{-1}$ (*2 s.f.*)

Comment 1

The unit mol^{-1} means *per mole of the equation as written* which, in this case, involves 2 mol of $NaHCO_3$.

Comment 2

ΔG is only slightly positive at 298 K, so the decomposition of sodium hydrogencarbonate is *almost* feasible at room temperature, but needs some heat input to become *actually* feasible. Again, this is in accord with everyday experience. The reaction is favoured by high temperatures and needs a temperature close to 400 K to become feasible (see the example on pages 23 and 24).

Notes

ΔG is positive, so at 298 K the reaction is *not feasible*. However, ΔG is not very large, so raising the temperature will allow the $-T\Delta S$ term to dominate. The reaction will become *feasible* at higher temperatures. The entropy change is positive since one gaseous mole forms for the loss of half a mole of solid.

Entropy in physical changes

In many physical changes, there is an increase or decrease in order (and hence in entropy). The most obvious of such changes is the melting of a solid or the boiling of a liquid, and these two cases are considered in the next two examples.

Melting (fusion)

When ice melts at constant temperature (0 °C) and constant pressure (100 kPa), the mixture of ice and water that results has *no spontaneous tendency* either to solidify (become all ice) or to liquefy (become all water) unless external influences (addition or removal of heat) are involved. If left in a system which prevents a flow of heat (e.g. a thermos flask) an ice/water mixture will remain at 0 °C as long as both ice and water are present. If a little heat leaks in, a little of the ice absorbs this heat and melts. The new equilibrium has slightly different amounts of ice and water present, but the temperature remains at 0 °C.

This is an example of a system at equilibrium that has no tendency to move spontaneously in one direction or the other.

Since it is the sign of ΔG that determines the tendency of a system to change, it can be concluded that:

For a system at equilibrium, $\Delta G = 0$

Applying this criterion, $\Delta G = 0$, to the equation $\Delta G^\ominus = \Delta H^\ominus - T\Delta S^\ominus$ leads to:

$$\Delta H^\ominus = T\Delta S^\ominus$$

So, for melting (*fusion*), $\Delta_{fus}S^\ominus = \dfrac{\Delta_{fus}H^\ominus}{T_{fus}}$

Boiling (vaporisation)

Arguments similar to the ones used for melting also apply here.

When water boils at constant temperature (100 °C) and constant pressure (100 kPa), the mixture of water and steam that results has *no spontaneous tendency* either to liquefy (become all water) or to vaporise (become all steam) unless external influences (addition or removal of heat) are involved. Left in a system which prevents a flow of heat, a water/steam mixture will remain at 100 °C as long as both water and steam are present. If a little heat leaks out, a little of the steam condenses and replaces the lost heat. The new equilibrium will have slightly different amounts of water and steam present, but the temperature will remain at 100 °C.

This is also a system at equilibrium that has no tendency to move spontaneously in one direction or the other.

Again, it is the sign of ΔG that determines the tendency of a system to change, so it can be concluded that:

for boiling (*vaporisation*), $\Delta_{vap}S^\ominus = \dfrac{\Delta_{vap}H^\ominus}{T_{vap}}$

The following examples illustrate the entropy changes that accompany the melting of ice and the boiling of water.

Example

Calculate the entropy change that accompanies the melting of ice. The **enthalpy of fusion** of ice is 6.0 kJ mol^{-1}.

Method

The reaction in question is:

$$H_2O(s) = H_2O(l)$$

This is an equilibrium change of state at fixed temperature, so $\Delta G = 0$.

$$\Delta_{fus}S^\ominus = \dfrac{\Delta_{fus}H^\ominus}{T_{fus}}$$

Answer

$$\Delta_{fus}H^\ominus = +6.0 \text{ kJ mol}^{-1}$$
$$T_{fus} = 273 \text{ K}$$

Therefore $\quad \Delta_{fus}S^\ominus = \dfrac{+6.0 \times 10^3 \text{ J mol}^{-1}}{273 \text{ K}} = +22 \text{ J K}^{-1}\text{ mol}^{-1}$

Comment

ΔS is positive, as is to be expected when a solid melts to form a liquid. The only difficulty worth noting is the need to convert enthalpy (in kJ mol^{-1}) into J mol^{-1} in order to give entropy in J K^{-1} mol^{-1}.

Example

Calculate the entropy change that accompanies the boiling of water.
The enthalpy of vaporisation of water is +44.0 kJ mol^{-1}.

Method

The reaction in question is:

$$H_2O(l) = H_2O(g)$$

This is an equilibrium change of state at fixed temperature, so $\Delta G = 0$.

$$\Delta_{vap}S^\ominus = \frac{\Delta_{vap}H^\ominus}{T_{vap}}$$

Answer

$$\Delta_{vap}H^\ominus = +44.0 \text{ kJ mol}^{-1}$$
$$T_{vap} = 373 \text{ K}$$

Therefore
$$\Delta_{vap}S^\ominus = \frac{+44.0 \times 10^3 \text{ J mol}^{-1}}{373 \text{ K}} = +118 \text{ J K}^{-1} \text{ mol}^{-1}$$

Comment

The entropy change here has a much more positive value than that
for a solid melting to form a liquid. The increase in entropy is associated
with the creation of one mole of disordered vapour molecules from
one mole of relatively ordered liquid molecules.

Calculation of the temperature at which a reaction becomes feasible

A reaction that is not feasible at one temperature may become feasible if the
temperature is raised (or lowered). The temperature at which feasibility is just
achieved is the temperature at which there is no spontaneous tendency for the
reaction to go either one way or the other. At this temperature, ΔG becomes
zero and

$$\Delta H^\ominus = T\Delta S^\ominus$$

So, for feasibility,
$$T = \frac{\Delta H^\ominus}{\Delta S^\ominus}$$

The temperature of feasibility can then be calculated if ΔH and ΔS are both
known. An illustration of such a calculation is given in the example shown
below.

Example

Calculate the temperature at which the thermal decomposition of sodium
hydrogencarbonate becomes feasible. The standard enthalpy and entropy
changes for the decomposition are given in Table 10.

The reaction in question is:

$$2NaHCO_3(s) = Na_2CO_3(s) + H_2O(g) + CO_2(g)$$

Method

Let 2 mol of PCl_5 decompose to form x mol of each of PCl_3 and Cl_2

Reaction	$PCl_5(g) \rightleftharpoons PCl_3(g) + Cl_2(g)$		

Initial moles   2   0   0

Equilibrium moles $(2 - x)$ x x

Total moles of gas $= (2 - x) + x + x = (2 + x)$

Calculation

Equilibrium moles:

PCl_3 and Cl_2 number of moles (given) x $= 1.2$ mol

PCl_5 number of moles $(2 - x) = (2 - 1.2)$ $= 0.8$ mol

Total moles of gas $= 0.8 + 1.2 + 1.2 = 3.2$ mol

Partial pressures:

$$p(PCl_5) = \frac{(2 - x)}{(2 + x)} \times p_{tot} = \frac{0.8}{3.2} \times 665 = 166 \text{ kPa}$$

$$p(PCl_3) = p(Cl_2) = \frac{x}{(2 + x)} \times p_{tot} = \frac{1.2}{3.2} \times 665 = 249 \text{ kPa}$$

Hence $K_p = \dfrac{p(PCl_3) \times p(Cl_2)}{p(PCl_5)} = \dfrac{249 \text{ kPa} \times 249 \text{ kPa}}{166 \text{ kPa}} = 374 \text{ kPa}$

Comment

Because there are two molecules of product and only one molecule of reactant in the equilibrium equation, the units of K_p in this case are kPa.

A change in total pressure affects the *equilibrium partial pressures but not the equilibrium constant*. Since

$$K_p = \left(\frac{\text{mole fraction } PCl_3 \times \text{mole fraction } Cl_2}{\text{mole fraction } PCl_5} \right) \times p_{tot}$$

then, if the total pressure *increases*, the term in brackets falls in compensation, leaving K_p constant. This result comes about through an *increase* in the mole fraction of undissociated PCl_5 and a simultaneous decrease in the mole fractions of PCl_3 and Cl_2.

Example 2

1 mol of nitrogen and 3 mol of hydrogen are heated to a temperature of 700 K under a pressure of 5000 kPa (5.0 MPa). The equilibrium mixture under these conditions contains 0.8 mol of ammonia. Calculate the value of the equilibrium constant K_p for the formation of ammonia from its elements under these conditions.

Method

Determine the partial pressures of the three species present using the equilibrium equation below. Let x mol of nitrogen react.

Reaction $\qquad\qquad\qquad$ $N_2(g) + 3H_2(g) \rightleftharpoons 2NH_3(g)$

Initial moles $\qquad\qquad\qquad$ 1 $\qquad$ 3 $\qquad\qquad$ 0

Equilibrium moles $\qquad\qquad$ $(1 - x)$ $\;$ $3(1 - x)$ $\qquad$ $2x$

Total moles of gas $= (1 - x) + 3(1 - x) + 2x = (4 - 2x)$

Calculation

Equilibrium moles and partial pressures:

NH_3 number of moles (given) $\qquad$ $2x$ $\qquad$ $= 0.8$ $\qquad$ Therefore $x = 0.4$ mol

$$p(N_2) = \frac{(1 - x)}{(4 - 2x)} \times p_{tot} = \left(\frac{0.6}{3.2}\right) \times 5 \qquad = 0.94 \text{ MPa}$$

$$p(H_2) = \frac{3(1 - x)}{(4 - 2x)} \times p_{tot} = 3\left(\frac{0.6}{3.2}\right) \times 5 \qquad = 2.81 \text{ MPa}$$

$$p(NH_3) = \frac{2x}{(4 - 2x)} \times p_{tot} = \left(\frac{0.8}{3.2}\right) \times 5 \qquad = 1.25 \text{ MPa}$$

Hence $\qquad$ $K_p = \dfrac{p(NH_3)^2}{p(N_2) \times p(H_2)^3} = \dfrac{(1.25 \text{ MPa})^2}{0.94 \text{ MPa} \times (2.81 \text{ MPa})^3} = 0.075 \text{ MPa}^{-2}$

Comment

Because there are two molecules of product and four molecules of reactant in the equilibrium equation, the units of K_p here are MPa^{-2}.

The qualitative effects of changes in temperature, pressure and concentration on the position of equilibrium and the value of the equilibrium constant K_c or K_p

The effect of changes in reaction conditions can be predicted by using **Le Chatelier's principle.**

> ### Definition – Le Chatelier's principle
> A system at equilibrium reacts to oppose any change imposed upon it.

Once the effect on the position of equilibrium is known, it is relatively easy to deduce the effect that this will have on the value of the **equilibrium constant.**

Change in temperature

A change in temperature changes the values of the equilibrium constants K_c and K_p. According to Le Chatelier's principle, the constraint of higher temperature can be relieved if the equilibrium moves in the direction that *absorbs* the added heat, and thus acts to lower the temperature.

Exothermic reactions

In an **exothermic reaction**, heat is given out as the reaction procees. This evolution of heat tends to *raise* the temperature of the reaction mixture. An increase in temperature can be opposed by reaction in the direction which will absorb the added heat and so *decrease* the temperature. Thus, in an exothermic reaction, the equilibrium is displaced to the *left*, and the equilibrium mixture contains a *lower concentration of products*. The converse is true if temperature is *decreased*.

Consider the effect of a change in temperature on the exothermic equilibrium reaction:

$$H_2(g) + I_2(g) \rightleftharpoons 2HI(g) \qquad\qquad \Delta H^\ominus = -9.6 \text{ kJ mol}^{-1}$$

for which the following values of K_c have been found:

Temperature/K	Equilibrium constant K_c
298	794
500	160
700	54

K_c and K_p *decrease* with *increasing* temperature in an *exothermic* reaction.

Endothermic reactions

In an **endothermic reaction**, heat is being taken in as the reaction proceeds, This absorption of heat will tend to *lower* the temperature of the reaction mixture. An increase in temperature can be opposed by reaction in the direction which will absorb the added heat so as to *decrease* the temperature. Thus, in an endothermic reaction, the equilibrium is displaced to the *right* and the equilibrium is mixture contains a *higher concentration of products*. The converse is true if temperature is *increased*.

Consider the effect of a change in temperature on the endothermic reaction:

$$N_2(g) + O_2(g) \rightleftharpoons 2NO(g) \qquad\qquad \Delta H^\ominus = +180 \text{ kJ mol}^{-1}$$

for which the following values of K_c have been found:

Temperature/K	Equilibrium constant K_c
298	4×10^{-31}
500	5×10^{-13}
1500	1×10^{-5}

K_c and K_p *increase* with *increasing* temperature in an *endothermic* reaction.

Notes

Increased temperature always shifts the equilibrium in the **endothermic** direction.

Decreased temperature always shifts the equilibrium in the **exothermic** direction.

Values of K_c and K_p are **altered** by changes in temperature unless $\Delta H^\ominus = 0$.

The effects of changes in temperature on equilibria are summarised below.

ΔH for reaction	Change in temperature	Shift of equilibrium	Yield of product	Equilibrium constant
exothermic	increase	to the left	reduced	reduced
exothermic	decrease	to the right	increased	increased
endothermic	increase	to the right	increased	increased
endothermic	decrease	to the left	reduced	reduced

Change in pressure

Changes in total pressure have a significant effect on the composition of a mixture at equilibrium *only if the reaction involves gases.* The changes observed are due to changes in the concentrations of the reactants and products. At a fixed temperature, the values of K_c and K_p are not affected.

According to Le Chatelier's principle, the constraint of additional pressure can be relieved if the equilibrium moves in the direction of fewer gaseous moecules, which exert less pressure and thus lower the overall pressure.

- If there are more moles of gaseous reactant than there are moles of gaseous product, an increase in total pressure will displace the reaction to the right to oppose the increase in pressure.

The system responds by trying to reduce the pressure by reducing the number of moles of gas present, and more product is formed. The converse of this also applies.

- In general, under *increased pressure,* the system moves to the side of the equation containing fewer moles of gaseous species. The important question is: which side has the greater number of gaseous moles? This can be expressed as the difference in numbers of gaseous moles, Δn:

$$\Delta n = n \text{ (products)} - n \text{ (reactants)}$$

Notes

Increased pressure always shifts the equilibrium towards the side with fewer moles of gas.

Consider, for example, the reaction:

$$CH_4(g) + H_2O(g) \rightleftharpoons 3H_2(g) + CO(g)$$

For this reaction, the equilibrium constants are given by:

$$K_c = \frac{[H_2(g)^3][CO(g)]}{[CH_4(g)][H_2O(g)]} \quad \text{and} \quad K_p = \frac{p(H_2)^3 \, (pCO)}{p(CH_4) \, p(H_2O)}$$

In this example, $\Delta n = +2$; the reverse reaction is favoured if pressure is increased. The fact that the value of the equilibrium constant K_p does not alter when the total pressure changes, cannot be *predicted* without having further information. At this stage, all that has to be remembered is:

Although the **position** of equilibrium shifts, the **value** of K_p and that of K_c remain the same, whatever the pressure.

The effects of changes in pressure on equilibria are summarised below.

More gas molecules on	Change in pressure	Shift of equilibrium	Yield of product	Equilibrium constant
product side	increase	to the left	reduced	unchanged
product side	decrease	to the right	increased	unchanged
reactant side	increase	to the right	increased	unchanged
reactant side	decrease	to the left	reduced	unchanged

Change in concentration

At a given temperature, the value of the equilibrium constant K_c is fixed. If the concentration of any species involved in an equilibrium is changed, then the concentrations of other species will change so that the value of K_c remains constant.

If the concentration of a reactant is increased, or the concentration of a product is decreased, the equilibrium opposes the change and is displaced to the right, giving more product. The opposite happens if reactant is removed or product is added.

Consider the reaction:

$$CH_3COOC_2H_5(l) + H_2O(l) \rightleftharpoons CH_3COOH(l) + C_2H_5OH(l)$$

for which the equilibrium constant is

$$K_c = \frac{[CH_3COOH(l)][C_2H_5OH(l)]}{[CH_3COOC_2H_5(l)][H_2O(l)]}$$

If more water is added, the equilibrium position is displaced to the right in order to restore the original value of K_c, and the equilibrium yields of CH_3COOH and C_2H_5OH are increased. In terms of the equilibrium constant, an *increase* in $[H_2O]$ is countered by a *decrease* in $[CH_3COOC_2H_5]$ (because some of it reacts with the added water) and a simultanoues *increase* in both $[CH_3COOH]$ and $[C_2H_5OH]$ (since more of these are formed). Similarly, if *more* CH_3COOH or *more* C_2H_5OH is added, equilibrium amounts of $CH_3COOC_2H_5$ and H_2O are *increased*.

If the reactants or products are *gases*, a change in the *partial pressure* of any gaseous species is equialent to a change in the *concentration* of that species.

The effect of a catalyst on the equilibrium position and on the equilibrium constant

A catalyst does not affect the equilibrium constant, neither does it have any effect on the position of equilibrium in a chemical reaction. Hence, catalysts do not affect the yield of chemical processes.

All that a catalyst can do is to *speed up the attainment of equilibrium*: it does so by providing an *alternative path* for the reaction with a *lower activation energy* (see *Collins Student Support Materials: AS/A-Level year 1 – Organic and Relevant Physical Chemistry*, section 3.1.5.5).

3.1.11 Electrode potentials and electrochemical cells

3.1.11.1 Electrode potentials and cells

Electrode potentials

Half-equations, electron transfer, reduction and oxidation

Half-equations (which represent half-reactions) have already been discussed in *Collins Student Support Materials: AS/A-Level year 1 – Inorganic and Relevant Physical Chemistry,* section 3.1.7. Electron transfer leads to electron gain by one species and electron loss by another. Electron gain is called **reduction**, electron loss is called **oxidation**. All redox reactions can be expressed as the sum of two half-reactions – one involving electron gain, and the other electron loss.

The IUPAC convention for writing half-equations

By convention, *all* redox half-equations are written as *reductions* (electron addition). This is the convention of the *International Union of Pure and Applied Chemistry* (IUPAC). Two such half-equations, written conventionally, are shown below:

$$Fe^{3+}(aq) + 3e^- \rightarrow Fe(s)$$
$$\tfrac{1}{2}Br_2(l) + e^- \rightarrow Br^-(aq)$$

Electrochemical cells

Redox reactions can be studied electrically using an **electrochemical cell**. A cell contains two **electrodes** (electrical conductors) immersed in an **electrolyte** (an ionic conductor), either as an aqueous solution or as a molten salt. An electrode together with its associated electrolyte forms an **electrode compartment**. Sometimes both electrodes share the same compartment (Fig 2) but if two compartments with different electrolytes are used, they are connected by means of a **salt bridge** (Fig 3).

A salt bridge consists of an electrolyte solution (often a saturated solution of KCl or KNO_3 in agar jelly), which completes the electrical circuit. It enables the cell to work by allowing ions to move between the two compartments while keeping apart the different solutions in the two compartments. If a wire were to be used instead of a salt bridge, it would introduce two more electrodes into the circuit.

Electrode reactions

In an electrochemical cell, each electrode compartment supports its own half-reaction. Electrons released by the oxidation half-reaction in one compartment, for example:

$$Zn(s) \rightarrow Zn^{2+}(aq) + 2e^-$$

Notes
The reaction represented by a *half-equation* is called a *half-reaction.*

Notes
Remember: *adding* electrons *reduces* positive charge.

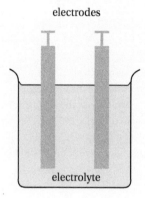
electrodes
electrolyte

Fig 2
A cell with a shared common electrolyte and a single compartment

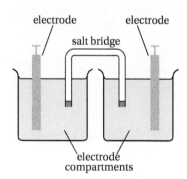
electrode — salt bridge — electrode
electrode compartments

Fig 3
A cell with two electrolytes in separate compartments connected by a salt bridge

are made available to drive the reduction half-reaction in the other compartment, for example:

$$Cu^{2+}(aq) + 2e^- \rightarrow Cu(s)$$

by flowing through the external circuit that completes electrical contact between the two compartments.

Electrodes

The simplest of all is the **metal electrode**, which comprises a metal in equilibrium with a solution of its ions. An example is the copper electrode which, depending on the position it occupies in the cell, is written as:

$$Cu(s) \,|\, Cu^{2+}(aq) \quad or \quad Cu^{2+}(aq) \,|\, Cu(s)$$

and is called the **electrode representation**. The vertical bars ($|$) shown above denote boundaries between different phases, which are themselves specified using state symbols.

The redox couple (e.g. Cu^{2+}/Cu) is a shorthand way of writing a reduction half-equation (e.g. $Cu^{2+}(aq) + 2e^- \rightarrow Cu(s)$). By convention, the oxidised species (without a state symbol) is written first, separated from the reduced species (also without a state symbol) by a forward slash (i.e. *Ox/Red*). This redox couple notation is not used in *cell diagrams* (see page 33) but is an economical way of specifying the half-equation in question.

A **gas electrode** consists of an inert metal (usually platinum) surrounded by a gas in equilibrium with a solution of its ions. The inert metal simply acts as either a source or a sink for electrons. The **hydrogen electrode**, whose special significance will be discussed shortly, is shown schematically in Fig 4 on page 34.

The hydrogen electrode corresponds to the redox couple H^+/H_2 and the electrode is denoted conventionally as:

Left electrode $Pt(s) \,|\, H_2(g) \,|\, H^+(aq) \quad or \quad H^+(aq) \,|\, H_2(g) \,|\, Pt(s)$ *Right electrode*

with *two* phase boundaries (solid–gas and gas–liquid) shown as vertical bars. The reduction half-reaction at this electrode is:

$$2H^+(aq) + 2e^- \rightarrow H_2(g)$$

A **redox electrode** is one at which two oxidation states of a given element undergo a reduction reaction at an inert metal surface, as in the case of the Fe^{3+}/Fe^{2+} couple:

Left $Pt(s) \,|\, Fe^{2+}(aq), Fe^{3+}(aq) \quad or \quad Fe^{3+}(aq), Fe^{2+}(aq) \,|\, Pt(s)$ *Right*

which has a reduction half-reaction, and a corresponding half-equation, written as:

$$Fe^{3+}(aq) + e^- \rightarrow Fe^{2+}(aq)$$

The redox couple MnO_4^-, H^+/Mn^{2+} would be denoted as an electrode by:

Left $Pt(s) \,|\, Mn^{2+}(aq), MnO_4^-(aq), H^+(aq) \quad or$

$MnO_4^-(aq), H^+(aq), Mn^{2+}(aq) \,|\, Pt(s)$ *Right*

corresponding to the reduction half-equation:

$$MnO_4^-(aq) + 8H^+(aq) + 5e^- \rightarrow Mn^{2+}(aq) + 4H_2O(l)$$

Notes

The electrode at which oxidation occurs is the *negative* electrode; the one at which reduction occurs is the *positive* electrode.

Essential Notes

Shorthand symbols like Cu^{2+}/Cu or Fe^{3+}/Fe are known as **redox couples**.
The standard electrode potentials of the reduction equations that they represent:
e.g. $Cu^{2+}(aq) + 2e^- \rightarrow Cu(s)$
and $Fe^{3+}(aq) + 3e^- \rightarrow Fe(s)$
are called **redox potentials**.

Notes

By convention, cell diagrams (see page 33) are always written with metal electrodes on the *outside*. For the electrode on the right, the electrode components are written in the same order as the couple (*Ox/Red*) with the reverse order (*Red/Ox*) used for the electrode on the left. Reading through a cell from left to right the order is *Red/Ox/Ox/Red* or ROOR.

Combining half-reactions into cell diagrams, and writing cell equations

A **cell diagram** is constructed by writing two electrodes back to back, and joining them with a salt bridge, conventionally denoted by two vertical bars.

$$Zn(s)\,|\,Zn^{2+}(aq)\,||\,Cu^{2+}(aq)\,|\,Cu(s)$$

There is a simple convention which helps when drawing a cell diagram with two electrode compartments joined together to form a cell:

Cell convention

The cell diagram is written with the more positive electrode (the one at which reduction occurs) shown as the right-hand electrode.

Essential Notes

The double bar signifies a junction across which there is no potential difference – usually a **salt bridge**. Sometimes it is shown as two vertical dashed bars.

Thus, in the zinc/copper cell above, the more positive electrode (Cu/Cu^{2+}) is correctly placed on the right-hand side of the cell diagram.

The **cell reaction** that corresponds to this cell diagram can be derived in one of two entirely equivalent ways, as follows:

Subtraction Method	Addition Method
• write the right-hand half-reaction as a *reduction*	• write the right-hand half-reaction as a *reduction*
• beneath it, write the left-hand half-reaction, also as a *reduction*	• beneath it, reverse the left-hand half-reaction, and write it as an *oxidation*
• *subtract* the left-hand (reduction) half-reaction from the right-hand one	• *add* the left-hand half-reaction (oxidation) to the right-hand one (reduction)
Right (red): $Cu^{2+}(aq) + 2e^- \rightarrow Cu(s)$	Right (red): $Cu^{2+}(aq) + 2e^- \rightarrow Cu(s)$
Left (red): $Zn^{2+}(aq) + 2e^- \rightarrow Zn(s)$	Left (ox): $Zn(s) \rightarrow Zn^{2+}(aq) + 2e^-$
Overall (by subtraction): $Cu^{2+}(aq) + Zn(s) \rightarrow Cu(s) + Zn^{2+}(aq)$	Overall (by addition): $Cu^{2+}(aq) + Zn(s) \rightarrow Cu(s) + Zn^{2+}(aq)$

Notes

As is self-evident, these two methods are entirely equivalent. Which one is chosen is simply a matter of personal preference.

This is the way the cell reaction actually goes; zinc does displace copper from solution.

Cell reactions

With the more positive electrode (the one at which reduction occurs) shown as the right-hand electrode, the cell reaction goes in the forward direction, as written.

Cell potential

In an electrochemical cell, a potential difference (*voltage*) is set up between the two electrodes. By convention, the left-hand electrode is more negative, relative to the right-hand electrode, which is more positive. Each electrode takes up its own characteristic potential and the overall **cell potential** is the difference between the two.

Factors that affect the cell potential

- *Cell current* – the true cell potential can be measured only under *zero-current* conditions, most commonly using a high-resistance digital voltmeter. The cell potential measured with zero current is called the *electromotive force* or the *EMF* (see below).

- *Cell concentration* – solution concentrations affect the cell potential. The *standard concentration* chosen is 1.00 mol dm^{-3}.

- *Cell temperature* – temperature affects the cell potential. The *standard temperature* chosen is 298 K.

- *Cell pressure* – pressure affects cell potentials, but not significantly unless a gas electrode is used. The *standard pressure* chosen is 100 kPa (1 bar) exactly.

The standard electromotive force (EMF)

Under standard conditions (zero current, 1.00 mol dm^{-3}, 298 K, 100 kPa) the cell potential is known as the *standard EMF of the cell*. The **cell EMF** does not have a special symbol.

> **Notes**
>
> If current is being drawn from the cell, the potential drops as the concentrations of the solutions in the cell change.

> **Notes**
>
> Measuring the EMF of an electrical cell is a required practical activity.

> **Definition**
>
> The **standard electromotive force (EMF)** is the potential difference between the electrodes of a standard electrochemical cell measured under zero-current conditions.

Essential Notes

The SI unit of electric potential difference is the volt (V).

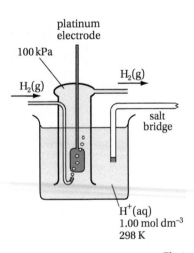

Fig 4
The hydrogen electrode. Hydrogen gas is bubbled over a platinum electrode establishing an equilibrium with $H^+(aq)$.

Standard Hydrogen Electrode (SHE)

It is impossible to measure the potential of a single electrode on its own, but if one electrode is assigned the value *zero*, then all other electrode potentials can be listed relative to this standard. The standard chosen is the hydrogen electrode operating under standard conditions.

It is called the **standard hydrogen electrode** (SHE).

$$\text{Pt(s)} \mid H_2(\text{g, 100 kPa)} \mid H^+(\text{aq, } 1.00 \text{ mol dm}^{-3}) \qquad E^{\ominus} = 0 \text{ at 298 K}$$

The basic form of this electrode was described earlier (see gas electrode on page 32) and is shown schematically in Fig 4. Under standard conditions, $E^{\ominus}(\text{SHE}) = 0$, by definition.

> **Definition**
>
> The potential, $E^{\ominus}(\text{SHE})$, of the **standard hydrogen electrode** is zero.

Standard electrode potential

The **standard electrode potential**, of any redox system or couple is found by measuring the potential of a cell with the SHE as the left-hand electrode and the unknown as the right-hand one.

All the variables affecting cell potential listed above *must* be kept under standard conditions when determining standard potentials.

Calculating $E^{\ominus}$ for an electrode from the *cell EMF*

The overall cell reaction is determined by subtracting the left-hand half-equation from the right-hand half-equation. The EMF is determined by subtracting the left-hand (L) electrode potential from the right-hand (R) one. Thus:

$$cell\ EMF = E^{\ominus}(R) - E^{\ominus}(L)$$

If the left-hand electrode is the SHE with $E^{\ominus} = 0$, then $E^{\ominus}_{right} = cell\ EMF$.

To ensure that the cell reaction goes in the forward direction, the less positive standard potential is put to the left and is subtracted from the right-hand one, giving a *cell EMF* > 0.

Some illustrations of the uses of electrode potentials and of *cell EMF* are shown in the following examples.

> *Example*
>
> Calculate the standard EMF of a cell with a silver electrode (Ag^+/Ag, $E^{\ominus} = +0.80$ V) and a copper electrode (Cu^{2+}/Cu, $E^{\ominus} = +0.34$ V), each in its own solution of 1.0 mol dm^{-3} silver(I) or 1.0 mol dm^{-3} copper(II) ions connected by a salt bridge.
>
> *Method*
>
> The more positive potential (Ag^+/Ag) is made the *right-hand electrode*.
>
> *Answer*
>
> $$cell\ EMF = E^{\ominus}_{right} - E^{\ominus}_{left}$$
>
> Therefore *cell EMF* = +0.80 − (+0.34) = +0.46 V

> *Example*
>
> Calculate the standard electrode potential of an unknown metal M which, in a standard solution of its ions M^{n+}(aq), gives a *cell EMF* of +0.48 V when acting as the right-hand electrode in a standard cell with tin ($E^{\ominus}$ (Sn^{2+}/Sn) = −0.14 V) as the left-hand electrode.
>
> *Method*
>
> Use:
>
> $$cell\ EMF = E^{\ominus}_{right} - E^{\ominus}_{left}$$

Notes

In the rest of this text the *subtraction method* (page 33) will be used, though it should be remembered that the *addition method* is equally valid.

Notes

Remember: R *minus* L.

Notes

If the cell potential is positive, the cell reaction goes forwards in the direction shown by the cell equation.

Answer

M^{n+}/M is the right-hand electrode and $E_{right}^{\ominus}$ = *cell EMF* + $E_{left}^{\ominus}$

$E^{\ominus} (M^{n+}/M) = +0.48 + (-0.14) = +0.34$ V

Example

Calculate the EMF of a cell with the *secondary standard* silver/silver chloride electrode ($E^{\ominus}$ = +0.22 V) as the left-hand electrode and the Fe^{2+}/Fe electrode ($E^{\ominus}$ = -0.44 V) as the right-hand one.

Method

Use:

$$cell\ EMF = E_{right}^{\ominus} - E_{left}^{\ominus}$$

Answer

cell EMF = -0.44 - 0.22 = -0.66 V

Comment

The *cell EMF* is negative, so the cell reaction will *not* go in the forward direction. Instead, the reverse reaction occurs:

$$2AgCl(s) + Fe(s) \rightarrow 2Ag(s) + Fe^{2+}(aq) + 2Cl^{-}(aq)$$

Notes

Because the SHE is complicated to set up and to use, a more convenient secondary standard electrode, already calibrated against the SHE, is used instead. So, an equally valid determination of an unknown standard potential may be made by measuring relative to a secondary standard electrode. Knowledge of secondary cells is not required in the AQA A-level specification.

Electrochemical series

Standard electrode potentials measured relative to the SHE can be placed in a list, either descending in voltage from the most positive (the convention adopted here) or ascending from the most negative. The choice is arbitrary, and it is necessary to accept and understand data presented in either format.

The data in Table 12 forms a part of the **electrochemical series**, with reduction half-reactions listed in order of decreasing electrode potential. In each case, the reactions listed can be thought of as an oxidising agent (on the left-hand side of a reduction half-equation) accepting electrons to form a reducing agent (on the right-hand side of a reduction half-equation).

- The strongest *oxidising* agents accept electrons easily and have more positive potentials.

- The strongest *reducing* agents lose electrons easily and have more negative potentials.

- Half-reactions with more positive potentials correspond to electron gain (**reduction**) reactions and go readily from left to right.

- Half-reactions with more negative potentials correspond to electron loss (**oxidation**) reactions and go readily from right to left.

Notes

Many of the helpful concepts for interpreting and using the electrochemical series, including those used in this book, depend on the order in which the data are presented. Flexibility of approach is needed to cope with any order of presentation.

Reduction half-reaction ox + e⁻ → red	$E^\ominus$/V
$F_2(g) + 2e^- \rightarrow 2F^-(aq)$	+2.87
$MnO_4^{2-}(aq) + 4H^+(aq) + 2e^- \rightarrow MnO_2(s) + 2H_2O(l)$	+1.55
$MnO_4^-(aq) + 8H^+(aq) + 5e^- \rightarrow Mn^{2+}(aq) + 4H_2O(l)$	+1.51
$Cl_2(g) + 2e^- \rightarrow 2Cl^-(aq)$	+1.36
$Cr_2O_7^{2-}(aq) + 14H^+(aq) + 6e^- \rightarrow 2Cr^{3+}(aq) + 7H_2O(l)$	+1.33
$Br_2(l) + 2e^- \rightarrow 2Br^-(aq)$	+1.09
$Ag^+(aq) + e^- \rightarrow Ag(s)$	+0.80
$Fe^{3+}(aq) + e^- \rightarrow Fe^{2+}(aq)$	+0.77
$MnO_4^-(aq) + e^- \rightarrow MnO_4^{2-}(aq)$	+0.56
$I_2(s) + 2e^- \rightarrow 2I^-(aq)$	+0.54
$Cu^{2+}(aq) + 2e^- \rightarrow Cu(s)$	+0.34
$Hg_2Cl_2(aq) + 2e^- \rightarrow 2Hg(l) + 2Cl^-(aq)$	+0.27
$AgCl(s) + e^- \rightarrow Ag(s) + Cl^-(aq)$	+0.22
$2H^+(aq) + 2e^- \rightarrow H_2(g)$	defined as 0
$Pb^{2+}(aq) + 2e^- \rightarrow Pb(s)$	−0.13
$Sn^{2+}(aq) + 2e^- \rightarrow Sn(s)$	−0.14
$V^{3+}(aq) + e^- \rightarrow V^{2+}(aq)$	−0.26
$Fe^{2+}(aq) + 2e^- \rightarrow Fe(s)$	−0.44
$Zn^{2+}(aq) + 2e^- \rightarrow Zn(s)$	−0.76
$Al^{3+}(aq) + 3e^- \rightarrow Al(s)$	−1.66
$Mg^{2+}(aq) + 2e^- \rightarrow Mg(s)$	−2.36
$Na^+(aq) + e^- \rightarrow Na(s)$	−2.71
$Ca^{2+}(aq) + 2e^- \rightarrow Ca(s)$	−2.87
$K^+(aq) + e^- \rightarrow K(s)$	−2.93
$Li^+(aq) + e^- \rightarrow Li(s)$	−3.05

Increasing Oxidising Power (left); *Increasing Reducing Power* (right)

Table 12
Standard electrode potentials at 298 K

The use of $E^\ominus$ values from the electrochemical series to predict the direction of simple redox reactions

The cell reaction occurs in the forwards direction of the cell equation if the right-hand electrode is the site of *reduction*. When this happens, the right-hand electrode is more positive than the left–hand electrode and $E^\ominus_{right} - E^\ominus_{left}$ is necessarily positive.

The direction of redox reactions

A cell reaction goes forwards, in the direction written, if, and only if, the corresponding *cell EMF* is positive.

In terms of the electrochemical series, a half-reaction with a more positive potential *oxidises* one with a more negative potential. In terms of the reduction equations in Table 12, the spontaneous direction of reaction (left to right, or right to left) involving pairs of half-reactions can be summarised as:

Cell EMF	Spontaneous direction
+ve	Forwards
–ve	Backwards

Essential Notes

Reminder:

Shorthand symbols like Zn^{2+}/Zn are known as *redox couples*.

The standard electrode potentials of the reduction equations that they represent:

e.g. $Zn^{2+}(aq) + 2e^- \rightarrow Zn(s)$

are called *redox potentials*.

The electrochemical series and *cell EMF*

Predictions about redox reactions can be made using the half-reactions in a table of standard redox potentials. Below are shown two methods of approach. Each of these methods is actually equivalent to the other, though at first sight they do appear to be different.

(i) Method 1 (the 'six-step' approach):

A series of six simple steps will always lead to the right answer. These are illustrated in Table 13, using the Ag^+/Ag and the Zn^{2+}/Zn redox couples as an example.

Table 13
Six steps in calculating *cell EMF*

Step 1

Consider the two half-reactions:

$Ag^+(aq) + e^- \rightarrow Ag(s)$ $\qquad\qquad E^\ominus = +0.80$ V

$Zn^{2+}(aq) + 2e^- \rightarrow Zn(s)$ $\qquad\qquad E^\ominus = -0.76$ V

Step 2

The half-equation with the *more positive* $E^\ominus$ value becomes the *positive electrode*; electrons arrive here from the external circuit, so:

$Ag^+(aq) + e^- \rightarrow Ag(s)$

The conventional half-reaction goes *forwards*.

Silver ions behave as the *oxidising* agent and are reduced.

Step 3

The half-equation with the *less positive* $E^\ominus$ value becomes the *negative electrode*; electrons leave this electrode and enter the external circuit, so:

$Zn(s) \rightarrow Zn^{2+}(aq) + 2e^-$

The conventional half-reaction goes *backwards*.

Zinc atoms behave as the *reducing* agent and are oxidised.

Step 4

The *overall equation* is obtained by adding the two new half-reactions (step 2 and step 3) so that *electrons cancel*:

$2Ag^+(aq) + Zn(s) \rightarrow Ag(s) + Zn^{2+}(aq)$

Note the need to double the Ag^+/Ag equation so that the electrons can cancel.

Step 5

The *cell representation* is obtained by placing the couple with the *more positive* $E^\ominus$ value on the right, with a salt bridge to separate it from the other couple:

$\overset{\ominus}{} Zn(s) \mid Zn^{2+}(aq) \mid\mid Ag^+(aq) \mid Ag(s) \overset{\oplus}{}$

The electrode polarities follow the rule: *positive on the right*.

Step 6

The **cell EMF** is obtained using *cell EMF* = $E^\ominus_{right} - E^\ominus_{left}$, so:

cell EMF = +0.80 − (−0.76) = +1.56 V

(ii) Method 2 (the 'outline sketch' approach):

Several of the stages listed in Table 13, particularly steps 1, 2, 3 and 5, can be represented as a diagram. This procedure is shown in Fig 5 and Fig 6 for the reaction between the Ag^+/Ag and the Zn^{2+}/Zn couples. A sketch based on the electrochemical series (Table 12) is drawn and the two selected half-reactions are identified (*step 1*, Table 13). The *higher* (more positive) couple is marked and the *lower* (less positive) one is marked (*steps 2 and 3*, Table 13). All other (intervening and outlying) half-equations and potentials in the electrochemical series can be ignored.

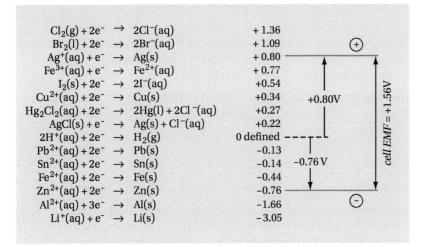

Fig 5
Determining the direction of spontaneous reaction and EMF for a redox couple

The graphical procedure is simplified by sketching the potential horizontally, so that it increases from left to right rather than upwards. Using the two chosen half-equations only, the diagram in Fig 5 is turned through 90° clockwise to fit the rule *'positive potential to the right'* (*step 5*, Table 13) and the reaction half-equations are omitted.

The result is an outline sketch, as in Fig 6, from which the cell representation can be written by simple inspection (*step 5*, Table 13). The direction of the cell reaction and the *cell* EMF (*steps 4 and 6*, Table 13) can then be determined using the rule *'R − L'*, or by reversing the oxidation (left-hand) equation and then adding.

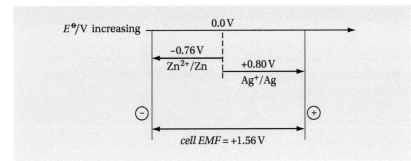

Fig 6
Outline sketch to determine the direction of spontaneous reaction and EMF for a redox couple

In the example shown, the cell representation is:

$$^{\ominus}Zn(s)\,|\,Zn^{2+}(aq)\,||\,Ag^+(aq)\,|\,Ag(s)^{\oplus}$$

the *spontaneous cell reaction* is:

Positive electrode:	$2Ag^+(aq) + 2e^- \rightarrow 2Ag(s)$
Negative electrode:	$Zn^{2+}(aq) + 2e^- \rightarrow Zn(s)$
Overall cell reaction (R – L):	$2Ag^+(aq) + Zn(s) \rightarrow Zn^{2+}(aq) + 2Ag(s)$

and the *cell EMF* is:

$$cell\ EMF = E^{\ominus}_{right} - E^{\ominus}_{left} = +0.80 - (-0.76) = +1.56\ V$$

These alternative approaches (*six steps* or *outline sketch*) are entirely equivalent and there is no need to try both. Instead, it is worth choosing the approach that comes most naturally and then practising to gain confidence in its use.

The determination of EMF and overall reaction is illustrated in the examples shown below. In the first example, both the *'six step'* and the *'basic sketch'* methods are used, followed by an illustration of each in the second and third examples.

Example

Use data from Table 12 to predict whether magnesium will reduce vanadium(III) ions to vanadium(II) ions. Write the representation of a standard cell in which reaction would occur and determine its EMF.

Method 1

Carry out the *six steps* in Table 13:

Step 1: $Mg^{2+}(g) + 2e^- \rightarrow Mg(s)$ $E^{\ominus}$ = –2.73 V
          $V^{3+}(aq) + 2e^- \rightarrow V^{2+}(aq)$ $E^{\ominus}$ = –0.26 V

Step 2: $V^{3+}(aq) + 2e^- \rightarrow V^{2+}(aq)$ $V^{3+}(aq)$ is the *oxidising* agent

Step 3: $Mg(s) \rightarrow Mg^{2+}(aq) + 2e^-$ Mg(s) is *oxidised*

Step 4: $2V^{3+}(aq) + Mg(s) \rightarrow Mg^{2+}(aq) + 2V^{2+}(aq)$

Step 5: $^{\ominus}Mg(s)\,|\,Mg^{2+}(aq)\,||\,V^{3+}(aq), V^{2+}(aq)\,|\,Pt(s)^{\oplus}$

Step 6:   *cell EMF* = –0.26 – (–2.37) = +2.11 V

Comment

The V^{3+}/V^{2+} couple reacts in the forward direction since it is more positive than the Mg^{2+}/Mg couple. Consequently, Mg will reduce V^{3+} to V^{2+}, itself being oxidised to Mg^{2+}.

Method 2

Draw the *outline sketch* as in Fig 6 on page 39:

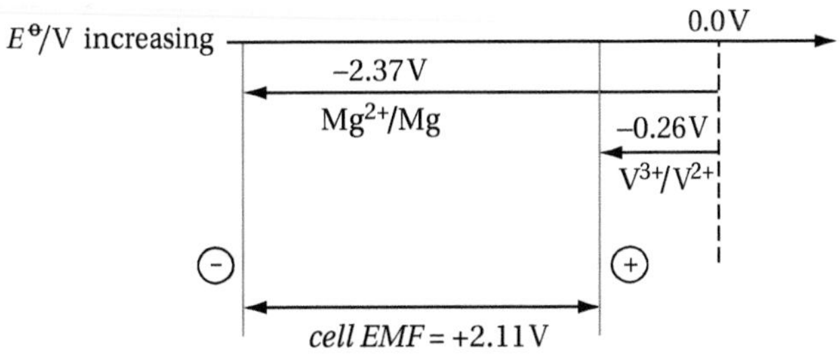

By inspection,
cell representation:

$$\ominus Mg(s)\,|\,Mg^{2+}(aq)\,||\,V^{3+}(aq),\,V^{2+}(aq)\,|\,Pt(s)\oplus$$

cell reaction (R – L):

$$Mg(s) + 2V^{3+}(aq) \rightarrow Mg^{2+}(aq) + 2V^{2+}(aq)$$

cell EMF (R – L):

$$\text{cell EMF} = -0.26 - (-2.37) = +2.11 \text{ V}$$

Comment

The V^{3+}/V^{2+} couple reacts in the forward direction since it is more positive than the Mg^{2+}/Mg couple. Consequently, Mg will reduce V^{3+} to V^{2+}, itself being oxidised to Mg^{2+}.

Notes

A platinum electrode is used to make electrical connection to the V^{3+}/V^{2+} couple. The oxidised and reduced forms of vanadium are not separated by a phase boundary, so a comma in the cell representation replaces the solid bar.

Example

Use data from Table 12 to determine whether chlorine can oxidise Fe(II) ions to Fe(III) ions. Write the representation of a standard cell in which this reaction might occur and determine its EMF.

Method 1

Carry out the six steps in Table 13.

Step 1: $Cl_2(g) + 2e^- \rightarrow 2Cl^-(aq)$ $E^{\ominus} = +1.36$ V

$Fe^{3+}(aq) + e^- \rightarrow Fe^{2+}(aq)$ $E^{\ominus} = +0.77$ V

Step 2: $Cl_2 + 2e^- \rightarrow 2Cl^-$ Cl_2 is the *oxidising* agent

Step 3: $Fe^{2+} \rightarrow Fe^{3+} + e^-$ Fe^{2+} is *oxidised*

Step 4: $2Fe^{2+}(aq) + Cl_2(g) \rightarrow 2Fe^{3+}(aq) + 2Cl^-(aq)$

Step 5: $\ominus Pt(s)\,|\,Fe^{2+}(aq),\,Fe^{3+}(aq)\,||\,Cl^-(aq)\,|\,Cl_2(g)\,|\,Pt(s)\oplus$

Step 6: *cell EMF* $= +1.36 - (+0.77) = +0.59$ V

In general terms, any such reaction could be written as:

$$R_{ox} + L_{red} \rightleftharpoons L_{ox} + R_{red}$$

where R and L refer to right-hand and left-hand cells, respectively, with the subscripts denoting the oxidised and reduced species in each half-cell.

Table 14
The effects of conditions on
cell EMF

	Increase in				
	$R_{ox}=[Fe^{3+}]$	$L_{red}=[V^{2+}]$	$L_{ox}=[V^{3+}]$	$R_{red}=[Fe^{2+}]$	Temperature
Effect on *cell EMF*	increase	increase	decrease	decrease	decrease

Notes
Altering cell concentrations or the temperature affects both $E^{\ominus}$ and EMF values, but the resulting changes are relatively small.

Changes in conditions have predictable effects on the *cell EMF*. However, the changes in EMF caused by moderate changes in concentration or temperature are rarely very large. For example, an increase in temperature of 10 °C in the iron/vanadium cell above will lower the EMF from 1.03 V to 1.01 V.

3.1.11.2 Commercial applications of electrochemical cells

Specially adapted electrochemical cells are commonly used as commercial sources of electrical energy. Such cells can be divided into three main categories: **non-rechargeable cells** (irreversible), **rechargeable cells** (reversible) and **fuel cells**:

Notes
Fuel cells are charged by the oxidation of a continuous supply of fuel, **not** by an electric current.

Definition

A *non-rechargeable (irreversible)* **cell** is not intended to be recharged by an electric current.

A *rechargeable (reversible)* **cell** is specifically designed to be recharged by an electric current.

A *fuel cell* generates electricity from the continuous oxidation of an external source of fuel.

Electrochemical cells that are used to provide a convenient source of electrical power are commonly called 'batteries', though the term 'battery' originates from an array of more than one cell joined in series to others.

Practical cells

Non-rechargeable cells

The sole function of a non-rechargeable (irreversible) cell is to provide current to an external circuit while discharging (galvanic action). Once discharged, non-rechargeable cells are discarded.

Characteristic features of some common non-rechargeable cells are shown in Table 15.

Notes
Knowledge of the details of these cells is **not** required in the AQA A-level specification; all that is needed is an ability to manipulate given equations and potentials to yield EMF values or vice-versa.

Table 15
Non-rechargeable cells.
The information shown below is for illustration only and does not form part of the AQA A-level specification.

Cell	Cell reactions	Cell EMF	Electrodes	Uses
Leclanché or Dry Cell	**Reduction:** $MnO_2(s) + H_2O(l) + e^- \rightarrow MnO(OH)(s) + OH^-(aq)$ **Oxidation:** $Zn(s) \rightarrow Zn^{2+}(aq) + 2e^-$ **Cell reaction:** $2MnO_2(s) + 2H_2O(l) + Zn(s) \rightarrow Zn^{2+}(aq)$ $\qquad\qquad + 2MnO(OH)(s) + 2OH^-(aq)$	1.5 V	**Reduction:** Graphite **Oxidation:** Zinc	Standard *'disposable'* cell
Alkaline cell	**Reduction:** $MnO_2(s) + H_2O(l) + e^- \rightarrow MnO(OH)(s) + OH^-(aq)$ **Oxidation:** $Zn(s) + 2OH^-(aq) \rightarrow Zn(OH)_2(s) + 2e^-$ **Cell reaction:** $2MnO_2(s) + 2H_2O(l) + Zn(s) \rightarrow Zn(OH)_2(s)$ $\qquad\qquad + 2MnO(OH)(s)$	1.54 V	**Reduction:** MnO_2 **Oxidation:** Zinc	*'long-life'* or *'high power'* cell *'Longer life'*: zinc casing not subject to acid attack. *'High power'*: provides a large steady current.
Mercury cell	**Reduction:** $HgO(s) + H_2O(l) + 2e^- \rightarrow Hg(l) + 2OH^-(aq)$ **Oxidation:** $Zn(s) + 2OH^-(aq) \rightarrow Zn(OH)_2(s) + 2e^-$ **Cell reaction:** $Zn(s) + HgO(s) + H_2O(l) \rightarrow Zn(OH)_2(s) + Hg(l)$ Mercury is very toxic, so careful disposal is needed at an official site rather than simply 'dumping'.	1.36 V	**Reduction:** HgO **Oxidation:** Zinc	Appliances with very low power needs. Small size, low current drain, long life, as in hearing-aids, digital watches and thermometers.
Lithium cell	**Reduction:** $Li^+ + MnO_2(s) + e^- \rightarrow LiMnO_2(s)$ **Oxidation:** $Li(s) \rightarrow Li^+ + e^-$ **Cell reaction:** $Li(s) + MnO_2(s) \rightarrow LiMnO_2(s)$ Lithium has the most negative of all reduction potentials, so batteries with large EMF values can be produced.	1.5 to 3 V	**Reduction:** MnO_2 **Oxidation:** Lithium	Digital cameras, calculators and watches. Very long life. Lithium iodide cells in pacemakers can last for up to 15 years.

Example

In the presence of ammonium chloride (commonly added to dry cells), the overall cell reaction is:

$$Zn(s) + 2MnO_2(s) + 4NH_4^+(aq) + 2OH^-(aq) \rightarrow$$
$$[Zn(NH_3)_4]^{2+}(aq) + 2MnO(OH)(s) + 2H_2O(l)$$

The $[Zn(NH_3)_4]^{2+}/Zn$ electrode has a standard potential of –1.03 V and the *cell EMF* is 1.5 V.

(a) Write the half-equation for the $[Zn(NH_3)_4]^{2+}/Zn$ electrode.

(b) Deduce the half-equation for the reaction at the other electrode (the right-hand electrode) of this dry cell.

(c) On the assumption that it operates under standard conditions, determine the standard electrode potential of the other electrode of this dry cell.

Answer

(a) This is a two-electron reduction with $[Zn(NH_3)_4]^{2+}(aq)$ as the oxidised species and $Zn(s)$ as the reduced one; so the reduction half-equation is:
$$[Zn(NH_3)_4]^{2+}(aq) + 2e^- \rightarrow Zn(s) + 4NH_3(aq)$$

(b) Adding this to the overall cell equation (given above), and cancelling terms, gives:
$$2MnO_2(s) + 4NH_4^+(aq) + 2OH^-(aq) + 2e^- \rightarrow$$
$$4NH_3(aq) + 2MnO(OH)(s) + 2H_2O(l)$$
and, allowing for the Brønsted–Lowry acid–base reaction of NH_4^+ and OH^- forming ammonia and water, further cancellation gives:
$$2MnO_2(s) + 2NH_4^+(aq) + 2e^- \rightarrow 2NH_3(aq) + 2MnO(OH)(s)$$

So, in its simplest form, the half-equation is:
$$MnO_2(s) + NH_4^+(aq) + e^- \rightarrow NH_3(aq) + MnO(OH)(s)$$

(c) Using $EMF = E_R - E_L$, with the $[Zn(NH_3)_4]^{2+}/Zn$ couple as the electrode on the left:
$$+1.5 = E_R - (-1.03), \text{ making } E_R = 1.5 - 1.03 = +0.47 \text{ V}$$

Comment

The ammonium chloride plays a pivotal role in ensuring that local concentrations of the battery components, those close to the electrodes, do not veer towards equilibrium conditions and thus decrease the EMF of the cell.

Notes

A *galvanic* cell is one with a positive EMF (> 0) in which the spontaneous forward cell reaction can be used to provide electric current to an external circuit.

An *electrolytic* cell is one with a negative EMF, which makes the forward cell reaction impossible, but which is then made possible by the flow of an electric current from an external source.

Rechargeable cells

These combine the two opposing functions of all reversible electrical cells, which are to provide current to an external circuit while discharging (*galvanic action*) and to use current from an external circuit while charging (*electrolytic action*). To be truly reversible, it is important that products resulting from both galvanic action and electrolytic action are not dispersed in the cell electrolyte but (being insoluble) remain attached to the cell electrodes.

Table 16

Rechargeable cells.

Except for the lithium ion cell, which is in section 3.1.11.2 of the AQA Specification, the information in Table 16 is for illustration only.

Cell	Cell reactions	Cell EMF	Electrodes	Uses
Lead–acid	**Reduction on discharge:** $PbO_2(s) + 3H^+(aq) + HSO_4^-(aq) + 2e^-$ $\xrightleftharpoons[\text{charge}]{\text{discharge}} PbSO_4(s) + 2H_2O(l)$ $\quad E^\ominus \approx +1.68\ V$ **Oxidation on discharge:** $PbSO_4(s) + H^+(aq) + 2e^-$ $\xrightleftharpoons[\text{discharge}]{\text{charge}} Pb(s) + HSO_4^-(aq)$ $\quad E^\ominus \approx -0.36\ V$ **Cell reaction:** $PbO_2(s) + Pb(s) + 2H^+(aq) + 2HSO_4^-(aq)$ $\xrightleftharpoons[\text{charge}]{\text{discharge}} 2PbSO_4(s) + 2H_2O(l)$ $\quad EMF \approx +2.04\ V$ In motor cars, these batteries are discharged at very high current when used to power a starter motor, and are then 'trickle' charged by an alternator when there is no current demand on the battery.	2.04 V	**Reduction:** PbO_2 **Oxidation:** Lead	The workhorse of secondary cells, in daily use in battery form in motorcars. So called 'sealed' car batteries do not need any maintenance. They are 'valve regulated', as 'sealed' could present very serious hazards.
Nickel–cadmium	**Reduction on discharge:** $NiO(OH)(s) + 2H_2O(l) + 2e^-$ $\xrightleftharpoons[\text{charge}]{\text{discharge}} Ni(OH)_2(s) + 2OH^-(aq)$ $\quad E^\ominus \approx +0.52\ V$ **Oxidation on discharge:** $Cd(OH)_2(s) + 2e^- \xrightleftharpoons[\text{discharge}]{\text{charge}} Cd(s) + 2OH^-(aq)$ $\quad E^\ominus \approx -0.88\ V$ **Cell reaction:** $NiO(OH)(s) + Cd(s) + 2H_2O(l)$ $\xrightleftharpoons[\text{charge}]{\text{discharge}} 2Ni(OH)_2(s) + Cd(OH)_2(s)$ $\quad EMF \approx +1.4\ V$	1.4 V	**Reduction:** $Ni(OH)_2$ **Oxidation:** Cadmium	All portable equipment needing high power, e.g. drills and chain saws.
Lithium ion	**Reduction on discharge:** $Li^+ + CoO_2 + e^- \xrightleftharpoons[\text{charge}]{\text{discharge}} LiCoO_2 \quad E^\ominus \approx +0.6\ V$ **Oxidation on discharge:** $Li^+ + e^- \xrightleftharpoons[\text{discharge}]{\text{charge}} Li \quad E^\ominus \approx -3.0\ V$ **Cell reaction:** $Li + CoO_2 \xrightleftharpoons[\text{charge}]{\text{discharge}} LiCoO_2 \quad EMF \approx +3.6\ V$ Lithium is very light and also has the highest negative standard electrode potential, so the EMF is very high.	3.6 V	**Reduction:** Lithium cobaltate **Oxidation:** Graphite	Fastest growing rechargeable battery system, in common use in mobile phones, iPods and lap-tops.

Characteristic features of some common rechargeable cells are shown in Table 16. The lithium cell is a good example of a rechargeable cell, commonly used in mobile phones and laptops. Details of the reactions can be found in Table 16, bottom row. The oxidation/reduction labels in the cell reactions refer to the discharge reactions. When charging, these reactions are reversed, replenishing the battery.

Fuel cells

Fuel cells use a supply of hydrogen or organic fuel, together with a supply of oxygen, to provide a source of electrical power. The simplest fuel cell uses energy from the reaction of hydrogen with oxygen to provide electrical power.

Fuel cells can replace the rather inefficient conversion (40%) into electricity from heat derived from combustion of hydrogen, with an electrochemical process that involves the continuous replenishment of fuel and oxygen at suitable electrodes to provide electrical power by a more direct method. Working at best attainable efficiencies, fuel cells can convert up to 85% or more of the chemical energy of combustion directly into electrical energy.

The alkaline **hydrogen–oxygen fuel cell** has two inert, porous, electrodes of nickel and nickel oxide (these permit the passage of the reactant gases as well as of the product, steam) immersed in a solution of hot aqueous potassium hydroxide, with gaseous hydrogen and oxygen bubbled into the cell continuously from opposite sides. The porous electrodes are impregnated with catalyst materials to speed up the electrode reactions and thus provide higher currents.

A schematic diagram is shown in Fig 7.

Notes

The current available from the cell depends on the rates of the electrode reactions, hence the advantage of having a catalyst present.

Fig 7
The alkaline hydrogen–oxygen fuel cell. Connections to an external circuit are shown as, in their absence, the hydrogen fuel cannot react with the oxygen. Electrons flowing in the external circuit are transferred from hydrogen to oxygen down the wire, rather than as a result of molecular collisions

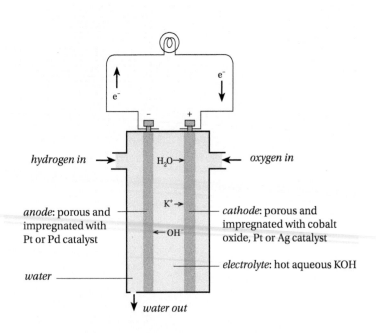

The electrode reactions are:

Reduction: $O_2(g) + 2H_2O(l) + 4e^- \rightarrow 4OH^-(aq)$ $E^\ominus/V = +0.40$

Oxidation: $2H_2(g) + 4OH^-(aq) \rightarrow 4H_2O(l) + 4e^-$ $E^\ominus/V = +0.83$

Overall: $2H_2(g) + O_2(g) \rightarrow 2H_2O(l)$ $EMF/V = +1.23$

The overall reaction is entirely equivalent to direct combustion of hydrogen in oxygen.

The favourable enthalpy of combustion of hydrogen and oxygen is converted directly into electrical energy. Fuel cells, therefore, provide an attractive means of energy production, in particular because they do not produce effluents that increase global warming.

Although this output EMF is based on standard conditions (298 K, 100 kPa), the hydrogen–oxygen fuel cell has only limited use under such conditions. The problem lies with the ability to produce appreciable electric current, and hence useable power. The diffusion rate of oxygen into the porous electrodes is not high at 298 K and, as a consequence, the current produced by the cell (which depends on the rate of production of electrons that, in turn, depends on the rate of conversion of gaseous oxygen to aqueous hydroxide ions) is low. To counteract this, it is usual to operate such cells at around 200 °C, with the higher temperature increasing the overall reaction rate (*kinetics*) and hence the available current.

Unfortunately, because the combustion of hydrogen is so very strongly exothermic, this increased temperature acts against the cell (*Le Chatelier*) and the EMF falls. An increased pressure of 20–40 bar can compensate for the increased temperature (*Le Chatelier*), and at 200 °C and 40 bar, an EMF of around 1.2 V can be achieved.

Hydrogen/oxygen fuel cells that operate at even higher temperatures (>500 °C) can achieve efficiencies in excess of 90%, but two notes of caution need to be expressed:

- Operating at a high temperature is sensible only if the heat generated can be used advantageously and not just dissipated. Lost heat seriously degrades the efficiency of the conversion process. Current thinking is moving towards **Combined Heat and Power (CHP)** systems, where excess heat can be used up, domestically, while power is being generated. Such systems are made even more cost-effective by feeding excess power, during 'idle' periods, into the electricity grid.

Notes

For clarity, it is useful to label the electrode half-reactions by their redox functions and then use the *addition method*.

Notes

The chemistry of the hydrogen–oxygen fuel cell illustrates how chemists go about solving the unremitting problems of slow rates and low yields. In this case, the 'yield' is electric current, which itself involves the rate of flow of electrons. So, to get useful power from a fuel cell, it is necessary to have fast reactions (increased temperature and use of catalysts) that provide a good yield of product (from an increased pressure of reactants). The chemical product is water, itself not of any commercial value, but the more water produced in a shorter time, the more electrical current that can flow, and hence the greater the electrical power output of the fuel cell.

- While oxygen (in the form of air) is in plentiful and cheap supply, hydrogen is not. Hydrogen needs to be manufactured and one of the common sources is the *steam reforming* reaction on natural gas:

$$CH_4(g) + H_2O(g) \rightarrow CO(g) + 3H_2(g)$$

with extra hydrogen available from the lower temperature reaction of carbon monoxide and steam:

$$CO(g) + H_2O(g) \rightarrow CO_2(g) + H_2(g)$$

Overall: $CH_4(g) + 2H_2O(g) \rightarrow CO_2(g) + 4H_2(g)$

Both of these reactions demand an input of energy, so hydrogen is not produced without an overall energy deficit. Moreover, this method leaves behind it a large carbon footprint, as one mole or more (the process is not 100% efficient) of carbon dioxide results for every four moles of hydrogen produced.

Overall, the 90% efficiency of conversion can outweigh the energy demands of hydrogen production (though not so much the carbon footprint), so hydrogen/oxygen fuel cells are firmly in place as a future source of supply of electrical energy. The power output of batteries of such high-temperature fuel cells can now exceed 10 megawatts, an amazing source of clean power, with only water as the ultimate 'pollutant'.

However, a strong note of caution is still needed. Fuel cells can provide a method of electricity generation that has a zero-carbon footprint only if the hydrogen needed is manufactured by some means that does not itself produce carbon dioxide. Such means of production are available, most notably by the electrolysis of acidified water, but these processes are only truly 'green' if the electricity needed for electrolysis is itself generated by carbon-neutral sources such as wind, tide, hydroelectric, or nuclear power. Before fuel cells can be viewed as 'clean', the carbon footprint of the production of source materials must be considered alongside the fact that the cells themselves produce only water as a product and do not, at first sight, appear to pollute.

3.1.12 Acids and bases

3.1.12.1 Brønsted–Lowry acid–base equilibria in aqueous solution

The Brønsted and Lowry definition of acids and bases states that:

> **Definition**
> A **Brønsted–Lowry acid** is a substance which **donates protons** in a reaction; it is a proton donor.
> A **Brønsted–Lowry base** is a substance which **accepts protons** in a reaction; it is a **proton acceptor**.

For example, consider the reaction between sodium carbonate and hydrochloric acid:

$$Na_2CO_3 + 2HCl \rightarrow 2NaCl + H_2O + CO_2$$

In ionic form, this can be written

$$CO_3^{2-}(aq) + 2H^+(aq) \rightarrow H_2O(l) + CO_2(g)$$

which shows that the HCl is a proton donor, i.e. a Brønsted-Lowry acid, and that the carbonate ion is a proton acceptor, i.e. a Brønsted-Lowry base.

Consider also the reaction that occurs when concentrated sulfuric acid and concentrated nitric acid are mixed:

$$H_2SO_4 + HNO_3 \rightleftharpoons H_2NO_3^+ + HSO_4^-$$

$\quad$ *acid* 1 $\quad$ *base* 1 $\quad$ *acid* 2 $\quad$ *base* 2

In this reaction, H_2SO_4 is behaving as a Brønsted-Lowry acid because it is donating a proton to HNO_3 which, rather surprisingly, is behaving as a Brønsted-Lowry base because it is accepting a proton. In the reverse direction, $H_2NO_3^+$ acts as the Brønsted-Lowry acid and HSO_4^- as the Brønsted-Lowry base.

> **Notes**
> It is perhaps surprising to find HNO_3 and HSO_4^- assuming the role of base, but that is exactly the role they adopt in this equilibrium.

The reaction shown above is an example of an **acid–base equilibrium**, which involves the transfer of a proton from H_2SO_4 to HNO_3. Addition of more H_2SO_4 or HNO_3 will shift the equilibrium to the right, producing more HSO_4^- and $H_2NO_3^+$, whereas addition of HSO_4^- or $H_2NO_3^+$ will move the equilibrium to the left.

Reactions of acids and bases in aqueous solution according to the Brønsted–Lowry scheme almost always involve the transfer of protons under equilibrium conditions:

$$H^+(aq) + B(aq) \rightleftharpoons BH^+(aq)$$

Here, the base B in aqueous solution accepts a proton (i.e. has a proton transferred to it). This happens, for example, when ammonia dissolves in water:

$$NH_3(g) + H_2O(l) \rightleftharpoons NH_4^+(aq) + OH^-(aq)$$

The acid (H_2O) transfers a proton to the base (NH_3) in the forward direction, and the acid (NH_4^+) transfers a proton to the base (OH^-) in the reverse direction.

The most fundamental proton-transfer equilibrium in water involves one water molecule, acting as a Brønsted–Lowry acid, donating a proton to another water molecule, acting as a Brønsted–Lowry base:

$$H_2O(l) + H_2O(l) \rightleftharpoons H_3O^+(aq) + OH^-(aq)$$

$$acid\ 1 \quad base\ 1 \quad acid\ 2 \quad base\ 2$$

3.1.12.2 Definition and determination of pH

The concentration of hydrogen ions in aqueous solution can vary over such a large range that it is convenient to express it on a logarithmic scale, called the pH scale.

pH is defined as:

> **Definition**
> $pH = -\log_{10}[H^+(aq)]$

Hydrogen ions are usually represented as $H^+(aq)$, or even simply as H^+ (especially in concentration terms such as $[H^+]$), rather than as the hydrated ion $H_3O^+(aq)$.

Measurement of pH

An indication of the pH of a solution can be obtained using a range of **indicators** (see this book, section 3.1.12.5) whose colour change with pH is known. The pH can conveniently be measured electronically using a pH meter which, for precise work, must first be calibrated against **buffer solutions** (see this book, section 3.1.12.6) whose pH is accurately known (usually at least two reference points are needed).

Strong acids and strong bases

In water, strong acids and strong bases are virtually completely dissociated into ions. At normal concentrations, acid or base molecules are *fully ionised*. There is only a limited number of strong acids in water. The most important of these are the *hydrohalic acids* (HCl, HBr and HI, but not HF which is a weak acid), nitric acid (HNO_3), sulfuric acid (H_2SO_4) and perchloric acid ($HClO_4$).

The vast majority of acids are weak (see this book, section 3.1.12.4) and, at normal concentrations in water, are only partially ionised. The same is true of most bases. In water, the only strong base normally encountered is the hydroxide ion (OH^-).

For strong acids in aqueous solution, the position of the proton-transfer equilibrium lies heavily to the right; proton transfer is virtually complete:

$$HCl(aq) + H_2O(l) \rightleftharpoons H_3O^+(aq) + Cl^-(aq)$$

Essential Notes

The square brackets around the symbol $H^+(aq)$ mean that the pH scale depends on hydrogen ion concentrations expressed in mol dm^{-3}.

Because of the logarithmic scale, pH values are normally quoted to 2 decimal places.

Notes

When present in high concentrations, even strong acids are not fully ionised due to interactions between the different species present in solution.

Essential Notes

It would be wrong to suggest that a strong acid must be 100% ionised. Any degree of ionisation appreciably above 50% entitles the acid to be classed as strong.

HCl is a Brønsted-Lowry acid; it donates a proton to water, and is a strong acid because the reaction goes practically to completion. Similarly for the strong acid sulfuric acid:

$$H_2SO_4(aq) + H_2O(l) \rightleftharpoons H_3O^+(aq) + HSO_4^-(aq)$$

In this case, the HSO_4^- ion is behaving as a base in the backward reaction, but it can also behave as an acid:

$$HSO_4^-(aq) + H_2O(l) \rightleftharpoons H_3O^+(aq) + SO_4^{2-}$$

Alkali metal hydroxides are strong bases; they dissociate fully when dissolved in a large volume of water and form aqueous hydroxide ions, OH^- (aq).

$$KOH(s) \xrightarrow{\text{water}} K^+(aq) + OH^-(aq)$$

The hydroxide ion is a Brønsted-Lowry base; it is a proton acceptor. Hydroxide ions react virtually completely with hydrogen ions to form water (*but* see this book, section 3.1.12.3).

$$OH^-(aq) + H_3O^+(aq) \rightleftharpoons 2H_2O(l)$$

which can be simplified to:

$$OH^-(aq) + H^+(aq) \rightleftharpoons H_2O(l)$$

Calculating the pH of an aqueous solution of a strong acid of known concentration

Strong acids can be regarded as fully ionised in dilute aqueous solution. Hence, the hydrogen ion concentration in a dilute solution of a monoprotic acid will be equal to the overall concentration of the acid in that solution.

Example

Calculate the pH of a 0.25 mol dm^{-3} solution of hydrochloric acid.

Method

Since the acid is fully ionised,
$[H^+]$ = overall concentration of HCl = 0.25 mol dm^{-3}

Answer

$$pH = -\log_{10}[H^+]$$

Hence pH = $-\log_{10} 0.25$ = 0.60

Calculating the concentration of an aqueous solution of a strong acid from its pH

Such a calculation is the reverse of the one given above.

Notes

Conductivity experiments can be used to differentiate between strong and weak acids or between strong and weak bases. With a simple circuit consisting of carbon electrodes, a bulb and a power supply, the brightness of the bulb indicates the extent of ionisation.

Essential Notes

A **monoprotic strong** acid (e.g. HCl) dissociates in water to produce one mole of protons per mole of acid. H_2SO_4 is a **diprotic** acid.

Notes

If the concentration of the acid is greater than 1 mol dm^{-3}, the pH is negative.

At a concentration of 1 mol dm^{-3}, the pH of a strong acid HX is 0.

At a concentration of 2.5 mol dm^{-3}, the pH of this acid is -0.40.

Notes

pH can never have very large negative values since these would demand very concentrated acid solutions (a pH of -1 implies a 10 mol dm^{-3} solution of H^+). Very concentrated solutions are impossible to obtain, because the concentration of H^+ is ultimately limited by the solubility of the acid in water as well as by ion association in such solutions.

Example

An aqueous solution of a strong monoprotic acid, HX, has a pH of 2.50.
Calculate the concentration of this acid.

Method

$$[H^+] = 10^{-pH}$$

So, convert pH into $[H^+]$ using inverse logarithms (antilogs).

Answer

$$pH = 2.50 = -\log_{10}[H^+]$$

Hence $[H^+]$ = antilog $(-2.50) = 10^{-2.5}$
$= 3.2 \times 10^{-3}$ mol dm^{-3}

Comment

Since the acid is a monoprotic strong acid, fully dissociated,
the concentration of HX = $[H^+] = 3.2 \times 10^{-3}$ mol dm^{-3}.

3.1.12.3 The ionic product of water, K_w

Water can act both as a Brønsted–Lowry acid (donating a proton) and as
a Brønsted–Lowry base (accepting a proton). As a result, both hydrogen
ions and hydroxide ions exist simultaneously in water according to the
equilibrium:

$$H_2O(l) \rightleftharpoons H^+(aq) + OH^-(aq)$$

and, because it is an equilibrium reaction, an equilibrium constant can be
derived:

$$K_c = \frac{[H^+(aq)][OH^-(aq)]}{[H_2O(l)]}$$

or, more simply: $K_c = \frac{[H^+][OH^-]}{[H_2O]}$ with units mol dm^{-3}.

However, water is only weakly dissociated, so the equilibrium position lies very
far over to the left (there are very few hydroxide or hydrogen ions present).
Consequently, the concentration of water $[H_2O(l)]$ can be taken to be constant
and its value incorporated into K_c. The new constant resulting from this is
called the **ionic product of water** and is given the symbol K_w. It is defined as
follows:

> **Definition**
>
> $K_w = [H^+][OH^-]$

Notes

The ionic constant of water
varies with temperature as
shown below:

T/K	K_w/mol^2 dm^{-6}
273	0.1×10^{-14}
293	0.7×10^{-14}
298	1.0×10^{-14}
303	1.5×10^{-14}
333	5.6×10^{-14}
373	51.3×10^{-14}

Notes

The pH of an iced drink
is about 7.5 (see values
of K_w above) and that of
a steaming mug of tea is
around 6.1, yet both are
neutral.

and has the units (mol dm^{-3})^2, or mol^2 dm^{-6}.

The value of K_w varies with temperature as shown above. K_w increases
as temperature increases because the ionic dissociation of water, which
involves the breaking of covalent bonds, requires an input of energy; it is an
endothermic process.

In neutral solution $[H^+] = [OH^-]$ always, so that $K_w = [H^+]^2$ and $[H^+] = K_w\sqrt{}$ at 298 K (when $K_w = 1.0 \times 10^{-14}$), $\sqrt{K_w} = 10^{-7}$, so pH = $-\log_{10}10^{-7} = 7.00$

Because K_w is temperature dependent and $[H^+] = [OH^-]$ *always* in neutral solutions, the pH of a neutral solution must vary with temperature. It is only at the single temperature 298 K, that the pH of pure water has, uniquely, the value of 7.

Using the value of K_w at 298 K, it can be seen that the pH scale from 0 to 14 spans a range of solutions from 1.0 mol dm^{-3} monoprotic strong acid (which has a pH of 0.00) to 1.0 mol dm^{-3} monoacidic strong base (which has a pH of 14.00).

Essential Notes

Strong acids react with strong bases to form salts (like NaCl or K_2SO_4), whose aqueous solutions are *always neutral*; at 298 K, salt solutions have a pH of about 7. Only solutions of salts of weak acids or weak bases (see this book, section 3.1.12.4) have a pH that differs from 7 at 298 K.

Calculating the pH of an aqueous solution of a strong base

Strong bases, like strong acids, are virtually completely ionised in water. So, using the value of K_w, we can calculate the pH of an aqueous solution of a strong base from its molar concentration in mol dm^{-3}.

Example

Calculate the pH at 298 K of a 0.15 mol dm^{-3} solution of sodium hydroxide. At 298 K, $K_w = 1.0 \times 10^{-14}$ mol^2 dm^{-6}.

Method

Since sodium hydroxide is fully ionised in aqueous solution, $[OH^-]$ = overall concentration of NaOH = 0.15 mol dm^{-3}.

Answer

$K_w = [H^+][OH^-] = 10^{-14}$ mol^2 dm^{-6}

Hence $[H^+] = \frac{K_w}{[OH^-]} = \frac{10^{-14}}{0.15} = 6.67 \times 10^{-14}$ mol dm^{-3}

and pH $= -\log_{10}(6.67 \times 10^{-14}) = 13.20$

Notes

At a concentration of 1 mol dm^{-3}, the pH of a strong base MOH is 14.

If the concentration of this base is greater than 1 mol dm^{-3}, the pH is greater than 14.

At a concentration of 2.5 mol dm^{-3}, the pH is 14.4.

Calculating the concentration of an aqueous solution of a strong base from its pH

Such calculations are the reverse of that given above.

Example

An aqueous solution of a strong **monoacidic base** MOH has a pH of 12.60. Calculate the concentration of this base.

Method

Convert pH into $[H^+]$ using inverse logarithms (antilogs), then use the value of K_w to convert $[H^+]$ into $[OH^-]$.

Notes

pH can never have a very large positive value, since this too would demand a very concentrated basic solution (a pH of 15 implies a 10 mol dm^{-3} solution of OH^- ions). Very concentrated solutions are impossible to obtain for the same solubility and association reasons that apply in concentrated acids.

Answer

pH = 12.60 $= -\log_{10}[H^+]$

Hence $[H^+]$ = antilog (–12.60) = $10^{-12.6}$ = 2.51×10^{-13} mol dm^{-3}

and $[OH^-] = \dfrac{K_w}{[H^+]} = \dfrac{10^{-14}}{2.51 \times 10^{-13}} = 4.0 \times 10^{-2}$ mol dm^{-3}

Comment

Since the base is a monoacidic strong base, fully dissociated, the concentration of MOH = $[OH^-]$ = 4.0×10^{-2} mol dm^{-3}.

3.1.12.4 Weak acids and bases, K_a for weak acids

Strong acids and **strong bases** ionise in water almost completely; **weak acids** and **weak bases** ionise in water only partially.

Weak acids

The extent to which an acid dissociates in water determines whether it is **weak** or **strong**. The strength of the acid is indicated by the position of the equilibrium established when the acid is dissolved in water. If the equilibrium lies to the right, the acid is a strong acid; if it lies to the left, the acid is weak.

Aqueous ethanoic acid is a commonly-encountered weak acid:

$$CH_3COOH(aq) \rightleftharpoons H^+(aq) + CH_3COO^-(aq)$$

The equilibrium here is well over to the left; the dissociation into ethanoate ions and hydrogen ions is only partial. Hence ethanoic acid is a weak acid.

Weak bases

Weak bases too are only partially ionised in aqueous solution. The strength of the base is indicated by the position of the equilibrium established when the base is dissolved in water. If the equilibrium lies to the right, the base is strong; if it lies to the left, the base is weak.

Aqueous ammonia is probably the most commonly-encountered weak base:

$$NH_3(aq) + H_2O(l) \rightleftharpoons NH_4^+(aq) + OH^-(aq)$$

The equilibrium here is well over to the left; the dissociation into ammonium ions and hydroxide ions is incomplete. Hence, ammonia is a weak base.

Notes

Investigation of pH changes when a weak acid reacts with a strong base and when a strong acid reacts with a weak base are required practical activities.

The acid dissociation constant K_a for weak acids
Definition of K_a
Consider a weak acid HA which dissociates only partially in aqueous solution:

$$HA(aq) \rightleftharpoons H^+(aq) + A^-(aq)$$

The acid dissociation constant, K_a, of acid HA, is defined as:

> **Definition**
>
> $$K_a = \frac{[H^+(aq)][A^-(aq)]}{[HA(aq)]}$$

The units in the expression for K_a can be cancelled:

$$\frac{\cancel{(mol\ dm^{-3})} \times (mol\ dm^{-3})}{\cancel{(mol\ dm^{-3})}} = mol\ dm^{-3}$$

The extent to which an acid dissociates in water is determined by the acid dissociation constant. The larger the value of the acid dissociation constant, the stronger the acid. Table 17 shows a number of acid dissociation constants at 298 K.

In general, acids with K_a much *smaller* than about 1 are classed as **weak acids** and those with K_a much *bigger* than 1 are classed as **strong acids.**

The value of K_a for HCl in Table 17 shows that when hydrogen chloride is dissolved in water, the equilibrium lies *very* far to the right, i.e. HCl is fully ionised in dilute aqueous solution; it is therefore a *very* strong acid.

As hydrofluoric and ethanoic acids have small acid dissociation constants, they are only slightly ionised in water. Their dilute aqueous solutions contain many undissociated acid molecules but few hydrogen ions; they are therefore weak acids. The value of the acid dissociation constant for hydrocyanic acid is extremely small, indicating that it is a *very* weak acid.

Approximate expression for K_a
For many weak acids, the expression for K_a:

$$K_a = \frac{[H^+(aq)][A^-(aq)]}{[HA(aq)]}$$

can be approximated if the extent to which HA dissociates is small. If so, then the concentration of acid in the denominator of the equation for K_a can quite reasonably be replaced by $[HA]_{tot}$ which is the *total original concentration* of HA. The hydrogen ion concentration can also be assumed to arise solely from dissociation of the acid and not at all from the ionisation of water. Thus, for a sufficiently weak acid, it is justifiable to write:

$$K_a \approx \frac{[H^+]^2}{[HA]_{tot}}$$

Acid	K_a/mol dm^{-3}
HCl	1.0×10^7
HNO$_3$	4.0×10^1
HF	5.6×10^{-4}
CH$_3$COOH	1.7×10^{-5}
HCN	4.9×10^{-10}

Table 17
Acid dissociation constants

Essential Notes
$[HA]_{tot} = [HA]_{eq} + [A^-]_{eq}$ so that

$[HA]_{eq} = [HA]_{tot} - [A^-]_{eq}$

where *eq* denotes the equilibrium concentration of the species in question.

As the acid is weak, and thus not strongly dissociated, $[A^-]_{eq}$ is small enough to be ignored, so that:

$[HA]_{eq} \approx [HA]_{tot}$ and, in addition:

$[H^+]_{eq} = [A^-]_{eq}$

This approximation can be used only in situations when the weak acid *alone* has been added to water. If the solution is then acidified, or if a base or the anion A⁻ is added, this approximation is no longer valid.

Calculating pH from K_a for a weak acid

The equation above can be used to find the hydrogen ion concentration, and hence the pH, of a weak acid. The following example illustrates the method.

Example

Calculate the pH of a 0.10 mol dm⁻³ solution of methanoic acid. K_a for methanoic acid is 3.6×10^{-4} mol dm⁻³.

Method

$$K_a = \frac{[H^+(aq)][A^-(aq)]}{[HA(aq)]}$$

$$\approx \frac{[H^+]^2}{[HA]_{tot}} \quad \textit{(weak-acid approximation)}$$

Therefore $[H^+] \approx \sqrt{K_a [HA]_{tot}}$

Answer

$K_a = 3.6 \times 10^{-4}$ mol dm⁻³ and $[HA]_{tot} = 0.10$ mol dm⁻³

Hence $[H^+] = \sqrt{3.6 \times 10^{-4} \text{ mol dm}^{-3} \times 0.10 \text{ mol dm}^{-3}}$

$$= \sqrt{3.6 \times 10^{-5} \text{ mol dm}^{-3}}$$

Thus $[H^+] = 6.0 \times 10^{-3}$ mol dm⁻³ and pH = 2.22

Comment

Increasing the concentration of a weak acid increases the hydrogen ion concentration and decreases the pH of the solution. Increasing K_a has the same effect. The weaker the acid, the higher the pH.

Definition of pK_a

Just as pH = $-\log_{10}[H^+]$, so a similar quantity pK_a can be defined:

Definition

$pK_a = -\log_{10}K_a$

A lower value of pK_a suggests a stronger acid (see Table 18); chloroethanoic acid is stronger than ethanoic acid which, in turn, is stronger than the ammonium ion.

Calculating pH from pK_a

These results can now be used in calculations of pH for solutions of weak acids and bases. Some examples are shown below.

Acid	pK_a
$HClO_2$	1.81
$CH_2ClCOOH$	2.85
HNO_2	3.37
HF	3.46
HCOOH	3.75
CH_3COOH	4.75
HOCl	7.53
HOBr	8.69
NH_4^+	9.25
HCN	9.31
$CH_3NH_3^+$	10.56

Table 18
pK_a values for selected acids

Example

Calculate the pH of a 0.025 mol dm^{-3} solution of nitric(III) (nitrous) acid. Use data given in Table 18 to work out your answer.

Method

$$HNO_2(aq) \rightleftharpoons H^+(aq) + NO_2^-(aq)$$

$$K_a = \frac{[H^+(aq)][NO_2^-(aq)]}{[HNO_2(aq)]} \approx \frac{[H^+]^2}{[HNO_2]_{tot}} \quad (\textit{weak-acid approximation})$$

Hence $[H^+] \approx \sqrt{K_a [HNO_2]_{tot}}$

Answer

$[HNO_2]_{tot} = 0.025$ mol dm^{-2} \qquad $pK_a = 3.37$

Hence $K_a = 10^{-3.37} = 4.266 \times 10^{-4}$ mol dm^{-3}

$[H^+] \approx \sqrt{4.266 \times 10^{-4} \times 0.025} = 0.003266$ mol dm^{-3}

Thus pH = 2.49

Example

Calculate the pH of a 0.50 mol dm^{-3} solution of ammonium chloride. Use data given in Table 18 in working out your answer.

Method

$$NH_4^+(aq) \rightleftharpoons H^+(aq) + NH_3(aq)$$

$$K_a = \frac{[H^+(aq)][NH_3(aq)]}{[NH_4^+(aq)]} \approx \frac{[H^+]^2}{[NH_4^+]_{tot}} \quad (\textit{weak-acid approximation})$$

Hence $[H^+] \approx \sqrt{K_a [NH_4^+]_{tot}}$

Answer

$[NH_4^+]_{tot} = 0.50$ mol dm^{-2} \qquad $pK_a = 9.25$

Hence $K_a = 10^{-9.25} = 5.623 \times 10^{-10}$ mol dm^{-3}

$[H^+] \approx \sqrt{5.623 \times 10^{-10} \times 0.50} = 1.677 \times 10^{-5}$ mol dm^{-3}

Thus pH = 4.78

Comment

The aqueous ammonium ion, formed when ammonium chloride dissolves in water, is a weak acid. The pH of the resulting solution is on the acid side of neutral.

3.1.12.5 pH curves, titrations and indicators

In analytical chemistry, the variation of pH during acid–base titrations can be used to determine the **equivalence point,** which corresponds to the mixing together of stoichiometrically equivalent amounts of acid and base. A plot of the pH of the solution being titrated against the volume of solution added is

Essential Notes

The **equivalence point** is also commonly called the **stoichiometric point**. Ideally, this will coincide with the **end-point** of the titration, which is a term that relates to colour change in an indicator (see later in this section) and should only be used in that context.

known as a pH curve. Some typical pH curves are shown in Figs 8 to 10; they represent the following titrations:

- Fig 8: *strong* base added to *strong* acid and *strong* base added to *weak* acid
- Fig 9: *strong* acid added to *strong* base and *strong* acid added to *weak* base
- Fig 10: *weak* base added to *weak* acid.

The shapes of pH curves in acid–base titrations

Figs 8 and 9 (*and their mirror images*) show the overall shapes predicted for **titration curves** in which 0.1 mol dm^{-3} solutions of various acids and bases are titrated together. The value of pKa for the weak acid and that for the protonated weak base have been chosen, respectively, as 4.75 (like the weak acid ethanoic acid) and 9.25 (like the ammonium ion, which is derived from the weak base ammonia).

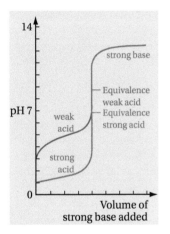

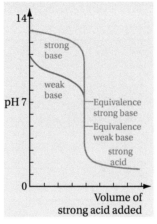

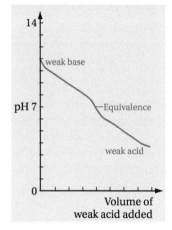

Fig 8
The titration curves of a strong acid (such as HCl) and of a weak acid (such as CH$_3$COOH) with a strong base (such as NaOH)

Fig 9
The titration curves of a strong base (such as NaOH) and of a weak base (such as NH$_3$) with a strong acid (such as HCl)

Fig 10
The titration curve of a weak base with a weak acid. The variation of pH with volume near the equivalence point is too gradual to allow for easy detection of equivalence.

Strong acid–strong base titrations

The two *strong acid–strong base* curves in Figs 8 and 9 are characteristic of the behaviour found for titrations of, for example, hydrochloric acid and sodium hydroxide. The reaction occurring is:

$$HCl(aq) + NaOH(aq) \rightarrow NaCl(aq) + H_2O(l)$$

Notes

An aqueous solution of the salt of a *weak acid* and a *strong base* is basic, with pH > 7.

An aqueous solution of the salt of a *strong acid* and a *weak base* is acidic, with pH < 7.

However, it should be recognised that the species indicated by the short-hand formulae HCl(aq), NaOH(aq) and NaCl(aq) are, in fact, fully dissociated in aqueous solution and exist only as individual H$^+$(aq), Cl$^-$(aq), Na$^+$(aq) and OH$^-$(aq) ions.

If HCl is titrated with NaOH, the initial pH is low (acidic region) and remains low while there is still acid present. Close to the equivalence point, the pH rises rapidly and, at the equivalence point, it reaches the pH of pure water (pH = 7), because the only major ions present [Na$^+$(aq) and Cl$^-$(aq)] have no effect on pH. After the equivalence point, the excess OH$^-$ ions present cause the pH to rise steeply again to a high value (basic region) where it remains, rising only slowly as more and more base is added.

The curve is traced out in reverse if NaOH is titrated with HCl.

Weak acid–strong base and strong acid–weak base titrations

At the equivalence point in a titration with a strong base (such as NaOH), the pH of a weak acid (such as CH$_3$COOH) lies above neutrality (pH = 7). At equivalence, the solution contains only the salt of a weak acid. The presence of a Brønsted-Lowry base (the CH$_3$COO$^-$ ion) means that the pH at equivalence is on the basic side of neutral (pH > 7). A similar argument explains why the pH at equivalence in the case of a weak base–strong acid titration lies on the acidic side of neutrality; the presence of a Brønsted-Lowry acid (NH$_4^+$ ions) means that the pH at equivalence is less than 7.

Four features are characteristic of titrations involving weak acids or bases with their strong counterparts, as can be seen in Figs 8, 9 and 11:

- **At equivalence**, the pH is not 7, but lies above this for weak acid titrations, or below this for weak base titrations. How to find the equivalence point from an experimental titration curve is shown in Fig 11.

- **Before equivalence**, the pH for weak acid titrations rises less steeply than with strong acid–strong base titrations.

- **After equivalence**, the change in pH follows exactly the trace of the appropriate strong acid–strong base titration. The strong acid or strong base present in excess beyond equivalence totally dominates the shape of the pH curve.

- **At the start of the titration**, for a weak acid (Fig 8), the pH rises more steeply than at the start of a strong acid–strong base titration, before flattening out and then rising towards equivalence. The flat portion is due to the formation of a **buffer solution** (see this book, section 3.1.12.6). For a weak base (Fig 9), the pH falls more steeply than at the start of a strong base–strong acid titration, before flattening out and then falling towards equivalence.

Concentrations and volumes of reaction for acids and bases

As the pH curves above have shown, there is an equivalence between a certain volume of an acid of one concentration and another volume of a base of a different concentration. The examples shown below illustrate how to calculate these volumes and concentrations.

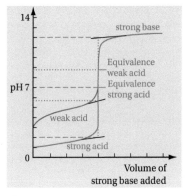

Fig 11
Graphical determination of the equivalence point. The equivalence point lies at the mid-point of the extrapolated vertical portion of the titration curve. The equivalence point for the strong acid is at pH 7.0, mid-way between pH 2.0 and 12.0. For the weak acid, this is at pH 8.9, mid-way between pH 5.8 and 12.0.

Example

28.4 cm^3 of a 0.22 mol dm^{-3} solution of Ba(OH)$_2$ are required to neutralise 25.0 cm^3 of a solution of HCl. Calculate the concentration of the HCl solution.

Notes

The species HCl, Ba(OH)$_2$ and BaCl$_2$ in aqueous solution are *fully dissociated* into ions and are *not* molecular species. Writing them as if they were molecular, establishes the stoichiometry more easily.

Method

The reaction occurring is:

$$2HCl(aq) + Ba(OH)_2(aq) \rightarrow BaCl_2(aq) + 2H_2O(l)$$

In *Collins Student Support Materials: AS/A-Level year 1 – Inorganic and Relevant Physical Chemistry,* section 3.1.2.5, stoichiometric calculations were carried out by determining:

(i)   the number of moles of the known substance

(ii)   the equivalence from the balanced equation, and hence

(iii)   the number of moles of the unknown substance.

Answer

(i)   moles Ba(OH)$_2$:   28.4 cm^3 of a 0.22 mol dm^{-3} solution

1000 cm^3 $\equiv$ 0.22 mol

$$28.4 \text{ cm}^3 \equiv \frac{0.22 \times 28.4}{1000} = 0.00625 \text{ mol}$$

(ii)   stoichiometry: 2 mol HCl $\equiv$ 1 mol Ba(OH)$_2$

(iii)   moles HCl: $2 \times 0.00625 = 0.0125$ mol in 25.0 cm^3

1000 cm^3 contain $0.0125 \times \dfrac{1000}{25.0} = 0.50$ mol

Hence the concentration of the HCl solution is 0.50 mol dm^{-3}.

Comment

It is easy to get the 2:1 ratio the wrong way round. If the base has two hydroxide ions as here then, for roughly equal volumes, they will be matched by close to a doubled concentration of monobasic acid – and vice-versa. It is always worth making this consistency check.

Notes

The species NaOH and Na$_2$C$_2$O$_4$ are *fully dissociated* into ions in aqueous solution and are *not* molecular species. However, writing them as if they were molecular, establishes the stoichiometry more easily.

Example

25 cm^3 of a solution of ethanedioic acid, H$_2$C$_2$O$_4$, is neutralised by 28.6 cm^3 of a 0.28 mol dm^{-3} solution of NaOH. Calculate the concentration of the ethanedioic acid solution.

Method

The reaction occurring is:

$$2NaOH(aq) + H_2C_2O_4(aq) \rightarrow Na_2C_2O_4(aq) + 2H_2O(l)$$

The three steps used in the previous example should be followed.

Answer

(i) moles NaOH: $28.6\ cm^3$ of a $0.28\ mol\ dm^{-3}$ solution

$1000\ cm^3 \equiv 0.28\ mol$

$$28.6\ cm^3 \equiv \frac{0.28 \times 28.6}{1000} = 0.0080\ mol$$

(ii) stoichiometry: $2\ mol\ NaOH \equiv 1\ mol\ H_2C_2O_4$

(iii) moles $H_2C_2O_4$: $\frac{1}{2} \times 0.0080 = 0.0040\ mol$ in $25\ cm^3$

$$1000\ cm^3\ contain\ 0.0040 \times \frac{1000}{25} = 0.16\ mol$$

Hence the concentration of the $H_2C_2O_4$ solution is $0.16\ mol\ dm^{-3}$.

Comment

For roughly equal volumes, the dibasic acid will be at approximately half the concentration of the NaOH.

pH calculations for strong acid–strong base titrations

To calculate pH during a titration, use the methods developed above. Two stages are involved in these calculations.

- Find the *number of moles present* of whichever component is *in excess*. If base is being added to acid, the acid will be in excess before equivalence, and the base after equivalence.

- Find the *total volume of solution*. Division of excess moles by total volume gives the *concentration* of the excess component, hence the pH of the solution.

The following examples illustrate this method.

Example

Calculate the pH in the titration of $10.0\ cm^3$ of a $0.15\ mol\ dm^{-3}$ solution of HCl at the point when $10.0\ cm^3$ of a $0.10\ mol\ dm^{-3}$ solution of NaOH have been added.

Method

Calculate the number of moles of acid originally present, the number of moles of base added, the number of moles of acid in excess, and the total volume of solution.

Answer

moles H^+ originally $= \dfrac{10}{1000} \times 0.15 \qquad = 0.0015\ mol$

moles OH^- added $= \dfrac{10}{1000} \times 0.10 \qquad = 0.0010\ mol$

Notes

It is essential to determine the total volume of the solution in order to find the *concentration* of hydrogen ions present. Forgetting the volume is a very easy mistake to make.

moles H^+ in excess	$= 0.0015 - 0.0010$	$= 0.0005$ mol
Total volume	$= 10.0 + 10.0$	$= 20.0$ cm^3
Hence $[H^+]$	$= \dfrac{0.0005 \times 1000}{20}$	$= 0.025$ mol dm^{-3}
pH	$= -\log_{10}[H^+]$	$= 1.60$

Comment

The acid is in excess, so the titration has not yet reached equivalence. Once the excess acid concentration has been determined, the calculation follows the examples shown earlier in this book, section 3.1.12.2.

Notes

The factor of 2 is used in finding the moles of H^+ added arises because sulfuric acid is a diprotic acid.

Also, it is essential to determine the total volume of the solution in order to find the concentration of hydrogen ions present. Forgetting the volume is a very easy mistake to make.

Example

Calculate the pH in the titration of 16.0 cm^3 of a 0.16 mol dm^{-3} solution of NaOH at the point when 12.0 cm^3 of a 0.10 mol dm^{-3} solution of H_2SO_4 have been added.

Method

The acid:base stoichiometry is 1:2

$$[H^+] = \frac{K_w}{[OH^-]_{xs}}$$

Answer

moles OH^- originally	$= \dfrac{16}{1000} \times 0.16$	$= 0.00256$ mol
moles H^+ added	$= 2 \times \dfrac{12}{1000} \times 0.10$	$= 0.00240$ mol
moles OH^- in excess	$= 0.00256 - 0.00240$	$= 0.00016$ mol
Total volume	$= 16.0 + 12.0$	$= 28.0$ cm^3
Thus $[OH^-]$	$= \dfrac{0.00016 \times 1000}{28}$	$= 0.00571$ mol dm^{-3}
Hence $[H^+]$	$= \dfrac{1.0 \times 10^{-14}}{0.00571}$	$= 1.75 \times 10^{-12}$ mol dm^{-3}
pH	$= -\log_{10}[H^+]$	$= 11.76$

Comment

The base is in excess, so the titration has not yet reached equivalence. Once the excess base concentration has been determined, the calculation follows the same procedure as shown earlier in this book, section 3.1.12.3.

pH calculations for weak acid–strong base titrations

The calculation of pH during these titrations involves using the methods developed earlier. The techniques used depend on how far the titration has progressed.

- **Before equivalence:** the relative proportions of weak acid and its anion present have to be determined and then used in the expression for K_a.

- **After equivalence:** the excess of strong base has to be found, together with the total volume of the solution, and the resulting concentration used to determine pH as in a strong base calculation (see the example earlier in this book, section 3.1.12.3). Each of these techniques is demonstrated in the following examples.

Notes

Before equivalence, the mixture of the weak acid and its anion behaves as a **buffer solution** (described later and illustrated in Fig 14 (page 70)).

Example

Calculate the pH in a titration when 10.0 cm^3 of a 0.10 mol dm^{-3} solution of NaOH has been added to 10.0 cm^3 of a 0.25 mol dm^{-3} solution of ethanoic acid ($K_a = 1.76 \times 10^5$ mol dm^{-3}).

Method

$$CH_3COOH(aq) \rightleftharpoons H^+ (aq) + CH_3COO^- (aq)$$

$$K_a = \frac{[H^+][CH_3COO^-]}{[CH_3COOH]}$$

So $[H^+] = K_a \times \dfrac{[CH_3COOH]}{[CH_3COO^-]}$

Answer

moles CH$_3$COOH originally	$= \dfrac{10.0}{1000} \times 0.25$	$= 0.0025$ mol
moles OH$^-$ added	$= \dfrac{10.0}{1000} \times 0.1$	$= 0.0010$ mol
moles CH$_3$COO$^-$ formed		$= 0.0010$ mol
moles CH$_3$COOH remaining	$= 0.0025 - 0.0010$	$= 0.0015$ mol
so [CH$_3$COOH]	$= 0.0015/V$	
and [CH$_3$COO$^-$]	$= 0.0010/V$	

Since both the ethanoic acid and the ethanoate ions exist together in the same overall volume, *concentration ratio = mole ratio*.

The volumes cancel so $[H^+] = K_a \times$ mole ratio $= 1.76 \times 10^{-5} \times \dfrac{0.0015}{0.0010}$

$$= 2.64 \times 10^{-5} \text{ mol dm}^{-3}$$

$$pH = -\log_{10}[H^+] \qquad = 4.58$$

Comment

The pH is in the acidic region. Note that it is only the *ratio* of concentrations of weak acid and its anion that matters here, so there is no need to consider the total volume since mole ratio $\equiv$ concentration ratio.

Example

Calculate the pH in a titration when 16.0 cm^3 of a 0.16 mol dm^{-3} solution of NaOH has been added to 12.0 cm^3 of a 0.20 mol dm^{-3} solution of ethanoic acid ($K_a = 1.76 \times 10^{-5}$ mol dm^{-3}).

Method

Since the strong base is in excess, the pH will be exactly the same as that in any equivalent strong acid–strong base titration.

Answer

In the example on page 64, 12.0 cm^3 of 0.10 mol dm^{-3} solution of H_2SO_4 were used, which is exactly equivalent to 12.0 cm^3 of a 0.20 mol dm^{-3} solution of ethanoic acid. Furthermore, the final volume in both titrations is the same.

Hence pH = 11.76

Comment

The pH is in the basic region. Note that it is *essential* here to take into account the total volume of the final solution.

The pH of a weak acid at half-equivalence

> **Definition**
>
> At **half-equivalence,** *exactly one-half of the equivalence volume of strong base has been added to the weak acid.*

Notes

This **half-equivalence** relationship provides a method for determining pK$_a$ by measuring the pH at half-equivalence.

This titration value has particular significance as it falls at the point where [HA] = [A$^-$] for the weak acid HA. The expression for K_a can be used to find the resulting pH through the equation:

$$[H^+] = K_a \times \frac{[HA]}{[A^-]}$$

The condition that the concentrations in numerator and denominator are the same (half-equivalence) means that:

$$[H^+] = K_a \quad \text{and} \quad pH = pK_a$$

This is shown graphically in Fig 14 (page 70).

> **Definition**
>
> At **half-equivalence,** *the pH of the solution of a weak acid has the same value as pK$_a$.*

Notes

The additional concentration of H$^+$ ions produced from HIn is too small to affect the final pH of the solution.

Indicators and their range of action

An acid–base indicator is a water-soluble, weak organic acid whose acid form (HIn) and base form (In$^-$) have different colours. At least one of the two colours needs to be intense, so that the addition of very little indicator (just a drop or two) will produce a clearly visible colour without appreciably disturbing the acid–base equilibrium to which it has been added.

At the equivalence-point of an acid–base titration, the pH changes very rapidly through several units of pH, and the indicator equilibrium:

$$HIn \rightleftharpoons H^+ + In^-$$

colour 1   *colour* 2

swings from almost *all HIn* to virtually *all In⁻* (or vice-versa). Thus, if a few drops of indicator have been added, there is an accompanying change from *colour* 1 to *colour* 2, or vice-versa, as the indicator equilibrium moves position under the influence of the changing concentration of H⁺ ions during the titration. This colour change is used to *indicate* the equivalence point (or, more correctly, the end-point – see below) of the titration.

The dominant species at low pH (when H⁺ ions are abundant) is the undissociated acid HIn. An abundance of H⁺ ions pushes the indicator equilibrium to the left. Conversely, at high pH, the anion In⁻ will dominate. Thus, at low pH the solution has the characteristic colour of HIn and at high pH it has the characteristic colour of In⁻. An indicator changes colour from only HIn to only In⁻ over quite a narrow range of pH.

Indicator	Colour 1 acid (HIn)	pH range	Colour 2 base (In⁻)
thymol blue	red	1.2 to 2.8	yellow
methyl orange	red	3.2 to 4.4	yellow
methyl red	red	4.8 to 6.0	yellow
litmus	red	5.0 to 8.0	blue
bromothymol blue	yellow	6.0 to 7.6	blue
phenol red	yellow	6.6 to 8.0	red
phenolphthalein	colourless	8.2 to 10.0	pink
alizarin yellow	yellow	10.0 to 12.0	red

Table 19
Characteristics of some common indicators

The end-point
The volume of titrant added to give a hydrogen ion concentration such that $[HIn] = [In^-]$ is called the **end-point** of the titration. The equivalence volume of titrant can be determined with precision when the indicator chosen has an end-point that coincides with the equivalence point of the titration.

The choice of indicator for a titration
An appropriate indicator for a given titration is best chosen by considering the specific pH curve for that titration. An indicator is appropriate if the rapid change of pH at equivalence overlaps the range of activity of the indicator. The reasons for the choice that is made can be seen by referring to Figs 12 and 13.

Fig 12 shows the pH curves for both a strong acid and a weak acid titrated with a strong base. In the strong acid case, the pH at equivalence is 7; for the weak acid, the pH at equivalence is shown as being 8.8 (as it is when a 0.1 mol dm⁻³ solution of ethanoic acid is titrated with a 0.1 mol dm⁻³ solution of sodium hydroxide).

Notes

The end-point (which refers to the *indicator*) should not be confused with the equivalence point (which refers to the *titration*).

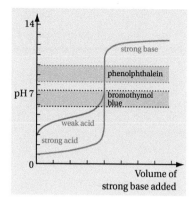

Fig 12
Titration of a strong acid and a weak acid with a strong base

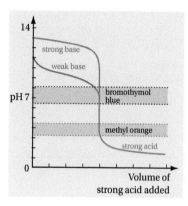

Fig 13
Titration of a strong base and a weak base with a strong acid

For the strong acid titration, bromothymol blue has a range that includes the equivalence value (pH = 7), so this would be a suitable indicator to choose. However, the change in pH is so steep and extended near equivalence that one drop of base is enough to swing the pH from 3 to 11. Any indicator in this extended range would be as good as bromothymol blue; as a result the choice of indicator is not very critical.

In the weak acid titration, an indicator with a range close to 7 would change colour before equivalence had been reached. With a pH at equivalence close to 9 in this titration, phenolphthalein would be a very suitable indicator.

Fig 13 shows the equivalent situation for a strong base and a weak base titrated with a strong acid. In the case of the strong base, the pH at equivalence is 7 as before; for the weak base, the pH at equivalence is shown as as being 5.2 (as it is when a 0.1 mol dm^{-3} aqueous solution of ammonia is titrated with a 0.1 mol dm^{-3} solution of hydrochloric acid).

Bromothymol blue has a range that includes pH = 7 but, once again, any indicator that changes colour in the pH range from 3 to 11 would be equally suitable. With a pH at equivalence close to 5 in this titration, methyl orange or methyl red would be very suitable indicators.

The pH changes at equivalence of several different titrations are shown in Table 20.

Acid	Base	pH range at equivalence	Choice of indicator (see Table 19)
HCl *strong*	NaOH *strong*	3 to 11	any from methyl orange downwards. The pH change here is over a very wide range
CH$_3$COOH *weak*	NaOH *strong*	7 to 11	any from phenol red downwards
HCl *strong*	NH$_3$ *weak*	3 to 7	methyl orange or methyl red
CH$_3$COOH *weak*	NH$_3$ *weak*	no sharp change	no suitable indicator, nor is the titration suitable with a pH meter since the pH variation shows no abrupt changes
HCl *strong*	Na$_2$CO$_3$ *weak*	2.5 to 5.5 *1st equivalence*	any from methyl orange to methyl red
		6.5 to 9.5 *2nd equivalence*	phenol red, thymol blue or phenolphthalein
H$_2$C$_2$O$_4$ *weak*	NaOH *strong*	1.5 to 3.5 *1st equivalence*	methyl orange or thymol blue; the pH change only just falls in the range of either indicator, so a titration using a pH meter is recommended
		5 to 11 *2nd equivalence*	any from bromothymol blue downwards

Table 20
The choice of an indicator for a given titration

3.1.12.6 Buffer action

Notes

A solution containing ethanoic acid and sodium ethanoate would make an *acidic buffer.*

A solution containing ammonium chloride and ammonia would make a *basic buffer.*

Definition

A **buffer solution** is one that is able to resist changes in pH when small amounts of acid or base are added. It is also able to maintain its pH in the face of dilution with water.

An **acidic buffer** is one that maintains a solution at a pH below 7 and, typically, consists of a weak acid and one of its salts (to provide the anion, which acts as a base).

A **basic buffer** is one that maintains a solution at a pH above 7 and typically consists of a weak base and one of its salts (to provide the cation, which acts as an acid).

Qualitative explanation of buffer action

Consider an acidic buffer consisting of a solution of ethanoic acid and sodium ethanoate. There is an equilibrium between the components of this solution:

$$CH_3COOH(aq) \rightleftharpoons CH_3COO^-(aq) + H^+(aq)$$

for which the equilibrium constant K_a, the acid dissociation constant, is

$$K_a = \frac{[CH_3COO^-][H^+]}{[CH_3COOH]}$$

which can be rearranged to give

$$[H^+] = K_a \frac{[CH_3COOH]}{[CH_3COO^-]}$$

so that the pH depends on the value of K_a and on the *ratio of the concentrations of the acid and anion (the base)*.

By looking at the equilibrium equation, or at the equilibrium expression for the hydrogen ion concentration, it is clear that:

- pH = pK_a when $[CH_3COOH] = [CH_3COO^-]$, so that the buffer pH will be on the acid side of neutrality.

- Addition of a *small* quantity of hydrogen ions will move the equilibrium to the left, causing ethanoate ions (in plentiful supply) to lower the excess hydrogen ion concentration by forming a little more ethanoic acid.

- Since the ethanoate ion and the ethanoic acid concentrations are both very large, this small increase from ethanoate ions to ethanoic acid

    $$A^- + H^+ \rightarrow HA$$

    will not change either the concentration of the acid or of the ion very much.

- As a result, since the ethanoate ion and ethanoic acid concentrations remain much the same, then so clearly does their ratio; the equations above show that the hydrogen ion concentration, and hence the pH, will also remain approximately constant.

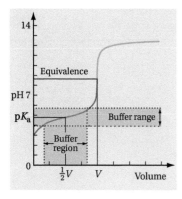

Fig 14
The buffer region for a typical weak acid such as ethanoic acid titrated with a strong base

Notes

Half-neutralised weak acids or weak bases, and also equimolar mixtures of the required components, are at the position of *half-equivalence*. (This situation was explored earlier in this book, in section 3.1.12.5.)

By direct analogy, the pH remains constant on the addition of small amounts of a base. In this case:

$$HA + OH^- \rightarrow A^- + H_2O$$

and the number of moles of OH^- added gives the number of moles of A^- formed as a result.

From the expression for $[H^+]$ above, it is also clear that dilution will not affect the pH since the *ratio of concentrations* will always be unchanged under dilution.

The presence of appreciable concentrations of both HA and A^- allows $[A^-]$ and $[HA]$ to remain virtually constant for small additions of acid or base and for the ratio of $[HA]$ to $[A^-]$ to remain constant on dilution. Hence $[H^+]$ and pH do not change.

The buffer region and buffer range

The small variation of pH when acid or base is added to a buffer can be understood with reference to the appropriate titration curve. Using as an example a weak acid (Fig 14), it is clear that in the region where the concentrations of acid and anion are similar (close to half-equivalence) the pH remains fairly constant (to within less than ±0.1 pH unit), even when small amounts of strong acid or base are added to the weak acid–anion mixture.

The region over which solutions can show buffer action is called the **buffer region** for the weak acid concerned. The **buffer range** is a term which describes the range of pH values in which a buffer can be prepared using a given weak acid. This range is usually taken to be $pK_a \pm 1$. For ethanoic acid, such a buffer range would be from about pH 3.7 to about pH 5.7.

Calculating the pH of a buffer solution
Acidic buffers
The pH of an acidic buffer solution can be calculated using the weak acid equation:

$$K_a = \frac{[H^+][A^-]}{[HA]}$$

leading to the expression

$$[H^+] = K_a \frac{[\text{weak acid}]}{[\text{salt}]}$$

where [salt] indicates the concentration of the anion of the weak acid present in solution.

A buffer solution may be made either by mixing solutions of the two buffer components (e.g. ethanoic acid and sodium ethanoate) or by partial neutralisation of the weak acid with a strong base (e.g. by adding about half of the stoichiometric amount of sodium hydroxide to a solution of ethanoic acid). In either case, an equimolar mix of components, or a half-neutralised weak acid or weak base, will have:

$$[H^+] = K_a \qquad \text{and hence} \qquad pH = pK_a$$

Two instances are shown in the following examples.

Example

Calculate the pH of a buffer made by mixing 14.0 cm^3 of a 2.0 mol dm^{-3} solution of ethanoic acid ($K_a = 1.74 \times 10^{-5}$ mol dm^{-3}) with 15.0 cm^3 of a solution of 1.50 mol dm^{-3} sodium ethanoate.

Method

Use the expression

$$[H^+] = K_a \frac{[CH_3COOH]}{[CH_3COO^-]}$$

Answer

1000 cm^3 of the acid solution contain 2.0 mol CH$_3$COOH

Thus
$$14.0 \text{ cm}^3 \equiv \frac{2.0 \times 14.0}{1000}$$
$$= 0.0280 \text{ mol CH}_3\text{COOH}$$

1000 cm^3 of the salt solution contains 1.50 mol CH$_3$COO$^-$

Thus
$$15.0 \text{ cm}^3 \equiv \frac{1.5 \times 15.0}{1000}$$
$$= 0.0225 \text{ mol CH}_3\text{COO}^-$$

and
$$[H^+] = K_a \times \frac{0.0280/V}{0.0225/V}$$
$$= 1.74 \times 10^{-5} \times 1.244$$
$$= 2.16 \times 10^{-5} \text{ mol dm}^{-3}$$

Hence   pH $= 4.66$

Comment

The pH is that of an *acidic buffer*. The absolute concentrations of the two components are not needed. It is the ratio of their concentrations that determines the buffer pH. Since the final volume of the solution is common to both components, the concentration ratio is the same as the mole ratio of the two components.

Example

Calculate the change in pH of the buffer solution in the above example after the addition of 10.0 cm^3 of a 0.10 mol dm^{-3} solution of HCl.

Method

The addition of acid shifts the equilibrium below to the *left*:
$$CH_3COOH(aq) \rightleftharpoons CH_3COO^-(aq) + H^+(aq)$$
with $[H^+] = K_a \dfrac{[CH_3COOH]}{[CH_3COO^-]}$

Decrease the number of moles of CH$_3$COO$^-$ by the number of moles of HCl added and increase the number of moles of CH$_3$COOH in proportion.

Answer

moles H^+ in 10.0 cm^3 of 0.10 mol dm^{-3} HCl $= \dfrac{0.10 \times 10.0}{1000} = 0.0010$ mol

Initial moles of CH_3COO^-	= 0.0225 mol
Thus new moles of CH_3COO^-	= (0.0225 − 0.0010) = 0.0215 mol
Initial moles of CH_3COOH	= 0.0280 mol
Thus new moles of CH_3COOH	= (0.0280 + 0.0010) = 0.0290 mol

and $[H^+]$ $= K_a \times \dfrac{0.0290}{0.0215}$

$\qquad\qquad = 1.74 \times 10^{-5} \times 1.349$ $= 2.347 \times 10^{-5}$ mol dm^{-3}

Hence pH = 4.63 when, previously, it was 4.66

The change in pH is −0.03 units.

Comment

Adding a substantial volume of a 0.10 mol dm^{-3} solution of HCl has only a slight effect on the pH, as is predicted for a buffer solution.

Applications of buffer solutions

In addition to laboratory use when standardising pH meter, mentioned above, buffer solutions have many additional laboratory uses in biological experiments on living systems. For example, the growth of bacteria for hospital tests is possible only in buffered systems; the waste products of growth, be they predominantly acidic or predominantly basic, can rapidly alter the pH of the growth medium to toxic levels, causing the bacteria to die.

A prime example of buffers in living systems is our own blood, which needs to be maintained at a pH close to 7.4 in healthy humans. The mechanism that maintains this pH is rather complex; it involves several aqueous buffer systems including the H_2CO_3/HCO_3^- and $H_2PO_4^-/HPO_4^{2-}$ pairs, and is enhanced by the buffering action of haemoglobin and other blood proteins.

The carbonic acid system acts through a linked system of chemical and physical equilibria:

$$H^+(aq) + HCO_3^-(aq) \rightleftharpoons H_2CO_3(aq) \rightleftharpoons H_2O(l) + CO_2(aq) \rightleftharpoons H_2O(l) + CO_2(g)$$

A straightforward application of Le Chatelier's principle down this sequence shows how an increase in acidity (a decrease in pH) in the blood is relieved by the formation of more aqueous H_2CO_3, relieved in turn by the production of more aqueous CO_2 and, finally, in the lungs, relieved by an increased breathing rate and the exhalation of more gaseous CO_2.

An inanimate example of a buffer system exists in the seas and oceans that surround us. These are maintained at a slightly alkaline level, at a pH of between 7.5 and 8.4, by a complicated system of buffers based on silicates and hydrogencarbonates. An enhanced knowledge of this system, which adjusts some of the level of CO_2 in our immediate environment, is clearly a vital tool in our understanding of global warming and climate change.

Notes

In a clinical condition called *acidosis*, the blood acquires too high a concentration of CO_2 and the pH drops. In order to expel the excess CO_2, rapid breathing and consequent discomfort result.

3.2 Inorganic chemistry

3.2.4 Properties of Period 3 elements and their oxides

The physical properties of the Period 3 elements (Na to Ar) were discussed in Collins Student Support Materials: *AS/A-Level year 1 – Inorganic and Relevant Physical Chemistry*, section 3.2.1.

This section considers the reactions of the elements with water and oxygen. The melting points of the oxides of the elements Na–S are also considered, together with the reactions of the oxides with water.

Reactions of the elements with water

Of the Period 3 elements, only sodium, magnesium and chlorine react with water. The reaction of chlorine with water was studied in Collins Student Support Materials: *AS/A-Level year 1 – Inorganic and Relevant Physical Chemistry,* section 3.2.3.2. Aluminium, silicon, phosphorus and sulfur do not react with water under normal conditions.

Sodium reacts violently with cold water. A piece of sodium added to cold water fizzes, skates over the surface of the water and becomes molten from the heat of the reaction. Hydrogen is evolved and this may catch fire and burn with a yellow flame (characteristic of sodium). At the end of the reaction, a colourless alkaline solution of sodium hydroxide remains:

$$2Na(s) + 2H_2O(l) \rightarrow 2Na^+(aq) + 2OH^-(aq) + H_2(g)$$

By contrast, as seen in Collins Student Support Materials: *AS/A-Level year 1 – Inorganic and Relevant Physical Chemistry,* section 3.2.2, magnesium reacts only very slowly with cold water but burns in steam when heated to give the white solid magnesium oxide and hydrogen:

$$Mg(s) + H_2O(g) \rightarrow MgO(s) + H_2(g)$$

Reactions of the elements with oxygen

The solid elements (Na to S) in Period 3 all burn in air or oxygen when ignited. Sodium burns with a yellow flame, forming the oxide:

$$2Na(s) + \tfrac{1}{2}O_2(g) \rightarrow Na_2O(s)$$

Magnesium, aluminium, silicon and phosphorus burn when ignited, emitting a very bright white light and white smoke of the oxides.

$$Mg(s) + \tfrac{1}{2}O_2(g) \rightarrow MgO(s)$$

$$2Al(s) + \tfrac{3}{2}O_2(g) \rightarrow Al_2O_3(s)$$

$$Si(s) + O_2(g) \rightarrow SiO_2(s)$$

$$P_4(s) + 5O_2(g) \rightarrow P_4O_{10}(s)$$

These reactions are very exothermic.

Sulfur burns with a blue flame, but much less vigorously than the elements above, to form the pungent, colourless gas sulfur dioxide:

$$S(s) + O_2(g) \rightarrow SO_2(g)$$

Notes

If there is a limited supply of oxygen, phosphorus also forms phosphorus(III) oxide, P_4O_6.

Notes

In an excess of pure oxygen, some sulfur trioxide is also formed.

Sulfur trioxide, SO_3, is produced industrially by the air oxidation of sulfur dioxide in the presence of vanadium(V) oxide catalyst.

$$SO_2 + O_2 \rightleftharpoons 2SO_3$$

Physical properties, structure and bonding of the oxides

The melting points, T_m, of the oxides are summarised in Table 21.

Table 21
Period 3 oxides

	Na_2O	MgO	Al_2O_3	SiO_2	P_4O_{10}	SO_2	SO_3
T_m/K	1548	3125	2345	1883	573	200	290
Bonding	ionic	ionic	ionic-covalent	covalent	covalent	covalent	covalent
Structure	lattice	lattice	lattice	macro-molecular	molecular	molecular	molecular

Notes

Sulfur also forms a higher oxide, SO_3, which can exist in several different forms, all of which have molecular structures with melting points below that of P_4O_{10}.

Ionic lattices are held together by strong electrostatic forces between ions, so that these lattices have high melting points. Macromolecular solids also have high melting points, because the atoms are held together by strong covalent bonds. Molecular solids involve weak intermolecular dipole–dipole or van der Waals' forces and have low melting points.

Reactions of the oxides

The reactions of the oxides with water and their structure and bonding are summarised in Table 22.

Table 22
Reactions of Period 3 oxides with water and approximate pH values of the resulting solutions

Reaction with water	pH	Structure and bonding in the oxide
$Na_2O(s) + H_2O(l) \rightarrow 2Na^+(aq) + 2OH^-(aq)$ very soluble; NaOH is a strong alkali	14	ionic lattice
$MgO(s) + H_2O(l) \rightarrow Mg^{2+}(aq) + 2OH^-(aq)$ sparingly soluble; $Mg(OH)_2$ is a weak alkali	9	ionic lattice
no reaction, Al_2O_3 insoluble	7	ionic–covalent lattice
no reaction, SiO_2 insoluble	7	macromolecular covalent
$P_4O_{10} + 6H_2O \rightarrow 4H_3PO_4$ very soluble; violent reaction; H_3PO_4 is a strong acid	0	molecular covalent
$SO_2 + H_2O \rightarrow H_2SO_3$ moderately soluble; H_2SO_3 is a weak acid	3	molecular covalent
$SO_3 + H_2O \rightarrow H_2SO_4$ very soluble; violent reaction; H_2SO_4 is a strong acid	0	molecular covalent

Notes

Phosphoric(V) acid is only a strong acid for the loss of its first proton to form $H_2PO_4^-$. Loss of all three protons to form PO_4^{3-} is very unlikely in water. PO_4^{3-} is formed when H_3PO_4 or P_4O_{10} react with an excess of base.

The trend across the period is:

alkaline oxides $\longrightarrow$ acidic oxides

The structure of the acids and anions formed when the oxides of phosphorus and sulfur react with water are shown in Table 23.

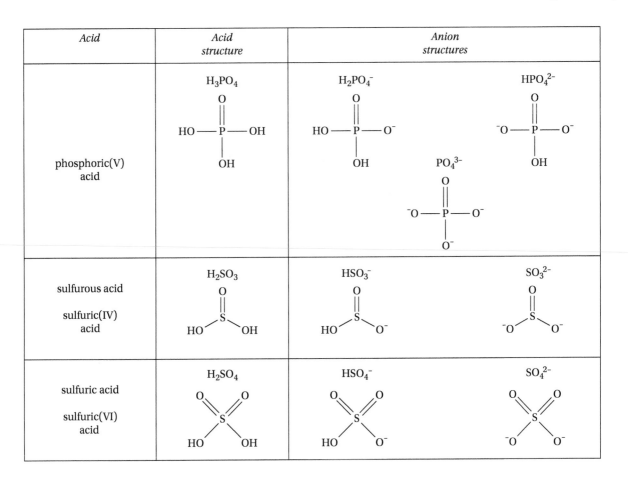

Acid	Acid structure	Anion structures		
phosphoric(V) acid	H_3PO_4	$H_2PO_4^-$	PO_4^{3-}	HPO_4^{2-}
sulfurous acid sulfuric(IV) acid	H_2SO_3	HSO_3^-		SO_3^{2-}
sulfuric acid sulfuric(VI) acid	H_2SO_4	HSO_4^-		SO_4^{2-}

Across the period, as the bonding in the oxide changes from ionic to molecular, the solutions of the oxides in water change from alkaline to acidic.

Thus the basic oxides of sodium and magnesium react with acids to form salts. The acidic oxides of phosphorus and sulfur, however, react with bases to form salts. In the middle of the period, aluminium oxide being insoluble in water does not affect pH; it is, however, capable of reacting with, and therefore dissolving in, both acids and bases. This behaviour is called **amphoterism** and aluminium oxide is said to be showing **amphoteric character**.

Table 23
Acids and anions formed when the oxides of phosphorus and sulfur react with water

Definition

An *amphoteric oxide is one which is capable of reacting with both acids and bases.*

Natural aluminium oxide (e.g. corundum), or laboratory samples which have been heated to red heat, are rather unreactive. Samples prepared at lower temperatures do, however, dissolve in acids to form salts and in alkalis to form aluminates. The amphoteric nature of aluminium hydroxide is discussed in this book, section 3.2.6.

Table 24 gives equations for some examples of these acid–base reactions.

Table 24
Reactions of oxides with acids
and bases

Oxide	Reaction with acid/base	
Na_2O	Basic, reacts with acid	$Na_2O + 2HCl \rightarrow 2Na^+ + 2Cl^- + H_2O$
MgO	Basic, reacts with acid	$MgO + 2HCl \rightarrow Mg^{2+} + 2Cl^- + H_2O$
Al_2O_3	Amphoteric, reacts with acid and with base	$Al_2O_3 + 6HCl \rightarrow 2Al^{3+} + 6Cl^- + 3H_2O$ $Al_2O_3 + 2NaOH + 3H_2O \rightarrow 2Na^+ + 2[Al(OH)_4]^-$
SiO_2	Acidic, reacts with base	$SiO_2 + 2NaOH \rightarrow 2Na^+ + SiO_3^{2-} + H_2O$
P_4O_{10}	Acidic, reacts with base	$P_4O_{10} + 12NaOH \rightarrow 12Na^+ + 4PO_4^{3-} + 6H_2O$
SO_2	Acidic, reacts with base	$SO_2 + 2NaOH \rightarrow 2Na^+ + SO_3^{2-} + H_2O$
SO_3	Acidic, reacts with base	$SO_3 + 2NaOH \rightarrow 2Na^+ + SO_4^{2-} + H_2O$

Notes

Either $[Al(OH)_4]^-$ or $[Al(OH)_6]^{3-}$ are accepted by examiners for the formula of the aluminate ion.

3.2.5 Transition metals

3.2.5.1 General properties of transition metals

The transition metals occupy the large central block of the Periodic Table, the **d block** (see *Collins Student Support Materials: AS/A-Level year 1 – Inorganic and Relevant Physical Chemistry,* section 3.2.1.1). However, in the present definition of a transition element, not all d-block elements are considered to be transition elements.

Notes

Scandium $[Ar]3d^14s^2$ is the first transition element, but is not included in the A-level specification.

Definition

A **transition element** *is an element having an incomplete d (or f) sub-level either in the element or in one of its common ions.*

The 3d sub-level can contain up to ten electrons; there are therefore ten elements in the first row of the d block from scandium to zinc. Titanium is the second element, the electron arrangement of which can be deduced from the Periodic Table as:

Ti $[Ar]3d^24s^2$

Notes

Note that $[Ar]4s^23d^2$ is equally acceptable as a representation.

This electron arrangement has an incomplete d sub-level, so that titanium is a transition element. Across the Periodic Table, the d sub-level is progressively filled until copper, which has the electron arrangement:

Cu $[Ar]3d^{10}4s^1$

This electron arrangement does not have an incomplete d sub-level. However, one of the common ions of copper is Cu^{2+}, which does have an electron arrangement with an incomplete d sub-level:

Cu^{2+} $[Ar]3d^9$

Notes

This electron arrangement is more stable than $[Ar]3d^94s^2$.

Copper is therefore a transition element.

After copper, the next element is zinc, which has the electron arrangement:

Zn $[Ar]3d^{10}4s^2$

and it forms only one common ion, which has the electron arrangement:

Zn^{2+} $[Ar]3d^{10}$

Neither in the element nor in its common ion does zinc have an incomplete d sub-level so that zinc is not classed as a transition element.

In working out the electron arrangement of a transition-metal ion, the *outer s electrons are always lost first* from the electron arrangement of the metal. Transition-metal ions and transition metals in compounds do not have any outer s electrons; it is the partly filled d sub-level which is responsible for the characteristic properties of the transition-metal ions in their compounds.

Characteristic properties of transition elements are:

- formation of complexes
- formation of coloured ions
- variable **oxidation states**
- catalytic activity.

Complex formation

Complex compounds contain a central atom surrounded by ions or molecules, both of which are called ligands.

Definition	Notes
A ligand is any atom, ion or molecule which can donate a pair of electrons to a metal ion.	Strictly, a lone pair donor is only a ligand when it is actually bonded to a metal.

The ability to donate a pair of electrons means that a **ligand**, a **Lewis base** and a **nucleophile** are equivalent.

When a complex compound is formed, ligands donate an electron pair to a metal ion to form a **co-ordinate bond** with the metal.

Definition	Notes
The number of atoms bonded to a metal ion is called the co-ordination number.	In a complex compound, the co-ordination number of the metal differs from its oxidation state.

In the hexaaquacopper(II) ion, each water molecule donates an electron pair (one of its lone pairs) to the copper(II) ion to form an octahedral complex:

$CuSO_4(s)$ $\xrightarrow{H_2O}$ $[Cu(H_2O)_6]^{2+}$
copper(II) sulfate   hexaaquacopper(II) ion
white   blue
    Co-ordination number of Cu = 6
    Oxidation state of Cu = +2

For ligands with only one donor atom, the co-ordination number is the number of ligands bonded to the metal ion; this is not true for multidentate ligands (see page 78).

It is important to remember that, while the hexaaqua complex is an ion, the bonds within the complex, i.e. the Cu—O and the O—H bonds, are covalent (see this book, section 3.2.6, for a diagram of the hexaaqua ion). Reactions of the complex ion involve the breaking of one or both of these types of bond.

3.2.5.2 Substitution reactions

A ligand such as water, which has only one atom that can donate a pair of electrons, and which consequently bonds through one atom only, is said to be **monodentate.** Monodentate ligands include:

H_2O, NH_3, Cl^-, OH^- and CN^-

Although several of these species have more than one lone pair of electrons, each ligand donates only one lone pair on co-ordination.

Ligands which contain two donor atoms, and which consequently are able to bond to a metal ion through two atoms, are called **bidentate.** Bidentate ligands include:

ethane-1,2-diamine (*ethylenediamine* or *en*), $H_2NCH_2CH_2NH_2$ (**Fig 15**), which bonds through *two* nitrogen atoms, and the ethanedioate (*oxalate*) ion, $C_2O_4^{2-}$ (**Fig 16**), which bonds through *two* oxygen atoms.

Fig 15
Ethane-1,2-diamine (*ethylenediamine*)

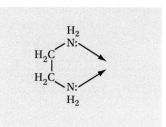

Fig 16
The ethanedioate (*oxalate*) ion

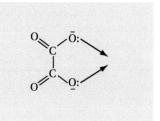

Some ligands contain many donor atoms and are said to be **multidentate.** Typical of these is the anion derived from bis[di(carboxymethyl)amino]ethane, commonly known as ethylenediaminetetraacetic acid or H_4EDTA. The anion **$EDTA^{4-}$** is able to bond to metal ions from six donor atoms.

The structure of the EDTA^{4-} anion is shown in Fig 17.

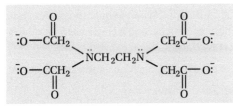

Fig 17
The structure of the EDTA^{4-} anion

EDTA uses the lone pairs on its six donor sites (4O and 2N) and forms 1:1 complexes with metal(II) ions, for example:

$$[Cu(H_2O)_6]^{2+} + EDTA^{4-} \rightarrow [Cu(EDTA)]^{2-} + 6H_2O$$

Blood contains a red iron(II) complex called haem. This complex is shown in Fig 18.

Four nitrogen atoms from a large organic molecule (called a porphyrin) are co-ordinately bonded in a plane, via their lone pairs, to the iron. The five-membered rings around the iron are shown with fixed positions for the double bonds but there is some delocalisation, similar to that in the benzene ring (see *Collins Student Support Materials: AS/A-Level year 2 – Inorganic and Relevant Physical Chemistry*, section 3.3.10.1). Octahedral co-ordination of the iron involves a fifth nitrogen atom, above the plane, from a protein called globin. The sixth position, below the plane, can be occupied by molecular oxygen co-ordinately bonded to the iron(II). This attachment enables oxygen to be transported in the blood. The structure is shown in Fig 18.

The affinity of this complex for carbon monoxide is much greater than its affinity for oxygen, and it is the formation of the very stable **carboxyhaemoglobin**, inhibiting the uptake of oxygen, that makes carbon monoxide so toxic.

Notes

EDTA can be used to estimate many metals volumetrically. An indicator is used which forms a weak complex with the metal ion. When all the free metal ion has been complexed with EDTA, the indicator is displaced from its weak metal complex and a new colour is seen at the end-point.

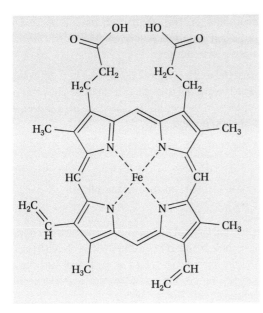

Fig 18
An iron(II) ion at the centre of a porphyrin ring in haem. Note that the haem structure is given on the AQA data sheet. Note that the dashed bonds in this structure represent co-ordinate bonds between a lone pair on an N atom and the central Fe atom.

Ligand substitution

Reactions in which one or more of the M—O bonds in the hexaaqua ion are broken will now be considered. When a water molecule is replaced by another ligand, a substitution reaction occurs; it is *nucleophilic substitution*, but is often just called **ligand substitution**.

For the purposes of substitution reactions, there are two types of ligand:

- *neutral* – uncharged molecules
- *anionic* – negatively-charged ligands.

Substitution by neutral ligands

Ammonia can act as a base in the *Brønsted–Lowry* sense (when it reacts with a proton) and also as a base in the *Lewis* sense (when it acts as a ligand).

A general equation can be written for the replacement of water molecules in an **aqua ion** by neutral ligands. For example, the replacement of H_2O by NH_3 can be written as:

$$[M(H_2O)_6]^{2+} + 6NH_3 \rightleftharpoons [M(NH_3)_6]^{2+} + 6H_2O$$

The equation above disguises the fact that what is written as a one-step equilibrium can be broken down into six steps, with only one water molecule being replaced in each step:

$$[M(H_2O)_6]^{2+} + NH_3 \rightleftharpoons [M(NH_3)(H_2O)_5]^{2+} + H_2O$$

$$[M(NH_3)(H_2O)_5]^{2+} + NH_3 \rightleftharpoons [M(NH_3)_2(H_2O)_4]^{2+} + H_2O$$

$$[M(NH_3)_2(H_2O)_4]^{2+} + NH_3 \rightleftharpoons [M(NH_3)_3(H_2O)_3]^{2+} + H_2O$$

$$[M(NH_3)_3(H_2O)_3]^{2+} + NH_3 \rightleftharpoons [M(NH_3)_4(H_2O)_2]^{2+} + H_2O$$

$$[M(NH_3)_4(H_2O)_2]^{2+} + NH_3 \rightleftharpoons [M(NH_3)_5(H_2O)]^{2+} + H_2O$$

$$[M(NH_3)_5(H_2O)]^{2+} + NH_3 \rightleftharpoons [M(NH_3)_6]^{2+} + H_2O$$

Essential Notes

Note the double 'm' in *ammine*; an *amine* is quite different.

At first sight, these equations look complicated, but they are in fact quite simple. The first equation starts with a hexaaqua ion and the last equation ends with a hexaammine ion. Note that, when water is bonded to a metal ion, it is called *aqua*, but when ammonia is bonded, it is called *ammine*, and there is no change in co-ordination number.

Ammonia is uncharged and has a similar size to water. Consequently, no change of shape is expected to occur during these substitution reactions. Thus, all the complexes in the equilibria above are **octahedral**, and there is no change in co-ordination number.

Substitution may be complete, for example:

$$[Co(H_2O)_6]^{2+} + 6NH_3 \rightleftharpoons [Co(NH_3)_6]^{2+} + 6H_2O$$

Substitution may also be incomplete, for example:

$$[Cu(H_2O)_6]^{2+} + 4NH_3 \rightleftharpoons [Cu(NH_3)_4(H_2O)_2]^{2+} + 4H_2O$$

With copper(II) ions, only four of the six water molecules on copper are replaced. When aqueous ammonia is added to a solution of a copper(II) salt; the first change observed is the formation of a *blue precipitate* of the hydroxide. This then dissolves when an excess of ammonia is added and a *deep-blue solution* of the tetraamminebisaquacopper(II) ion is formed.

$$[Cu(H_2O)_6]^{2+} + 4NH_3 \rightleftharpoons [Cu(NH_3)_4(H_2O)_2]^{2+} + 4H_2O$$
blue solution   deep-blue solution

Further substitution can be achieved when the concentration of ammonia is increased (e.g. by cooling the solution in ice and saturating it with ammonia gas, or by using liquid ammonia rather than aqueous ammonia), but in concentrated aqueous ammonia, the equilibrium position reached is that in which the dark-blue $[Cu(NH_3)_4(H_2O)_2]^{2+}$ ion is formed. This ion has four ammonia molecules in a square-planar arrangement around copper with water molecules occupying the other two octahedral positions above and below the plane. The bonds to water are longer and weaker than the bonds to ammonia.

Notes

A more sophisticated explanation for the lack of formation of the hexaammine, based on the Jahn–Teller effect, can be found in undergraduate texts.

Substitution by chloride ions

All anions are potentially capable of acting as ligands. For example, the very common ligand Cl^- has the electron arrangement of the noble gas argon; it has four lone pairs of electrons. When Cl^- bonds to a metal ion, one of these lone pairs forms a co-ordinate bond to the metal; the three remaining lone pairs on the chloride ion do not co-ordinate.

A good source of chloride ions is concentrated hydrochloric acid. Hydrogen chloride is very much more soluble in water than ionic chlorides such as sodium chloride, allowing high concentrations to be achieved; this is why concentrated hydrochloric acid is used to prepare complexes of chloride ions with transition metals.

Notes

Concentrated hydrochloric acid contains about $11 \ mol \ dm^{-3}$ of HCl and is virtually completely dissociated:
$$HCl + H_2O \rightarrow H_3O^+ + Cl^-$$

When a pink solution of a cobalt(II) salt in water is treated with an excess of concentrated hydrochloric acid, a deep-blue solution is formed:

$$[Co(H_2O)_6]^{2+} + 4Cl^- \rightleftharpoons [CoCl_4]^{2-} + 6H_2O$$
pink   blue
octahedral   tetrahedral

When the blue solution of $[CoCl_4]^{2-}$ is diluted with water, the pink colour returns. This behaviour can be understood in terms of an equilibrium which can be driven from left to right by high chloride ion concentrations and from right to left if the concentration of chloride ions is lowered by dilution. This marked change in colour is brought about not only by the change of ligand but, more significantly, also by the change in the co-ordination number of the cobalt ion. The shape of the ligands around the cobalt ion changes from octahedral to **tetrahedral** as hydrochloric acid is added.

Notes

Note that a high concentration of chloride ions is necessary to push the equilibrium over to the right-hand side; it is hard to achieve so high a concentration even with saturated aqueous sodium chloride.

Why then does the shape of the complex ion change? The chloride ligand is:

- negatively charged

- and large.

As the size of the ligands around a metal ion is increased, a point is reached when the electron charge-clouds around the ligands repel each other to such an extent that the octahedral structure becomes less stable than the tetrahedral one. In the tetrahedral structure, the ligands are approximately 109° apart and do not experience repulsive forces as great as in the octahedral arrangement, where the ligands are only 90° apart.

Thus, the general equation for the substitution reaction of a metal(II)-aqua ion with chloride ions is:

$$[M(H_2O)_6]^{2+} + 4Cl^- \rightleftharpoons [MCl_4]^{2-} + 6H_2O$$
octahedral   tetrahedral

Substitution by bidentate and multidentate ligands

Example

Deduce what will happen when concentrated hydrochloric acid is added to a solution of copper(II) sulfate.

Method

Recall the effect of ligand substitution on shape, co-ordination number, and colour in a complex ion.

Answer

The copper(II)-aqua ion is *octahedral*. When small water molecules are replaced by bigger chloride ions, a *tetrahedral* complex is likely to be formed. The *co-ordination number* will fall from 6 to 4. There will probably be a change of colour in the solution.

Comment

The blue solution of copper(II) ions turns yellow-green once an excess of hydrochloric acid has been added and the tetrachlorocuprate(II) ion is formed:

Equation:	$[Cu(H_2O)_6]^{2+} + 4Cl^- \rightleftharpoons [CuCl_4]^{2-} + 6H_2O$	
Colour:	blue	yellow-green
Shape:	octahedral	tetrahedral
Co-ordination number:	6	4

When the solution is diluted, the decreased concentration of chloride ions forces the equilibrium back to the hexaaqua ion and the solution turns blue again.

When a ligand with two donor atoms attacks a metal-aqua ion, two water molecules are replaced in the first instance. For example, a bidentate ligand such as ethane-1,2-diamine (*ethylenediamine* or *en*) reacts with a metal(II)-aqua ion according to the equation:

$$[M(H_2O)_6]^{2+} + H_2NCH_2CH_2NH_2 \rightleftharpoons [M(H_2O)_4(H_2NCH_2CH_2NH_2)]^{2+} + 2H_2O$$

As more ligand is added, further substitution can occur until all the water molecules have been replaced:

$$[M(H_2O)_4(H_2NCH_2CH_2NH_2)]^{2+} + H_2NCH_2CH_2NH_2 \rightleftharpoons [M(H_2O)_2(H_2NCH_2CH_2NH_2)_2]^{2+} + 2H_2O$$

$$[M(H_2O)_2(H_2NCH_2CH_2NH_2)_2]^{2+} + H_2NCH_2CH_2NH_2 \rightleftharpoons [M(H_2NCH_2CH_2NH_2)_3]^{2+} + 2H_2O$$

In these reactions with ethylenediamine, the equilibrium position lies well over to the right-hand side; the equilibrium constants for the formation of tris(ethylenediamine) complexes are of the order of 10^{20}. This large value means that the resulting complexes are much more stable than the aqua ions

Notes

If the size of the ligand is decreased to the smaller F^- ion, then octahedral complexes become common, e.g. AlF_6^{3-} (c.f. $AlCl_4^-$). If, however, the size of the ligand is increased to the larger I^- ion, octahedral complexes are not formed with any of the cations of the first three periods.

from which they were formed. Note that, because the donor atoms are small, there is no change in shape and the complexes remain octahedral.

Metal(III)-aqua ions react similarly, so that the equation for the reaction with an excess of ethane-1,2-diamine is:

$$[M(H_2O)_6]^{3+} + 3H_2NCH_2CH_2NH_2 \rightleftharpoons [M(H_2NCH_2CH_2NH_2)_3]^{3+} + 6H_2O$$

These metal(III)-tris(ethylenediamine) complexes are even more stable than the metal(II) complexes; the equilibrium constant for the above reaction is of the order of 10^{30}.

This extra stability is known as the **chelate effect**. Complexes in which bidentate or multidentate ligands bond to one metal ion only are known as chelates. The ligand forms a five- or six-membered ring with the metal ion, as shown in Fig 19.

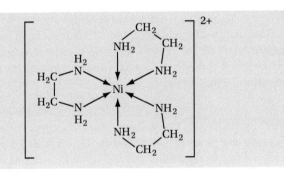

Chelating ethylenediamine

Chelating ligands can be present in complexes which also contain unidentate ligands as in, for example, the ion shown in Fig 20, *trans*-dichlorobis(ethylenediamine)cobalt(III).

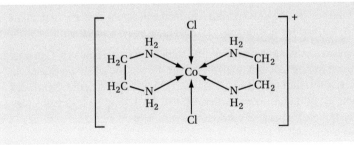

Notes

Hint: When drawing complex ions with *multidentate ligands*, first draw the six co-ordinate bonds arranged octahedrally around the central metal atom, *then* fit the ligands to these bonds.

Fig 20
Unidentate chloride ions as part of a bidentate chelate complex

Another common bidentate ligand is the di-anion ethanedioate, $C_2O_4^{2-}$. This ligand forms a very stable complex with iron(III) ions, $[Fe(C_2O_4)_3]^{3-}$, shown in Fig 21.

Fig 21
The $[Fe(C_2O_4)_3]^{3-}$ ion

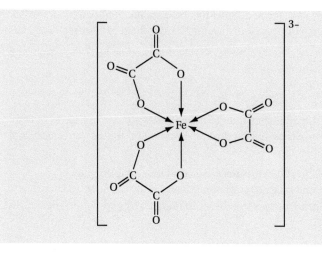

Multidentate ligands form even more stable complexes than do bidentate ligands. Haem in blood is an example taken from iron chemistry. The **hexadentate** ligand $EDTA^{4-}$ reacts with metal(II)-aqua ions to form a complex in which the ligand occupies all six octahedral sites around the metal ion, liberating six water molecules:

$$[M(H_2O)_6]^{2+} + EDTA^{4-} \rightleftharpoons [M(EDTA)]^{2-} + 6H_2O$$

The $EDTA^{4-}$ ligand (see Fig 17) finds many uses in analytical and industrial chemistry. The complexes formed are very stable, so that all the equilibria are completely over to the right-hand side and very few aqua ions are present. Reactions involving the metal-aqua ion are therefore not possible in the presence of an excess of $EDTA^{4-}$, so that metal ions can be kept in solution (sequestered), even when anions which normally cause precipitation, such as OH^- or CO_3^{2-}, are added.

Thus, addition of an excess of the sodium salt of $EDTA^{4-}$ to aqueous copper(II) sulfate, followed by aqueous sodium hydroxide, does not give any precipitate of copper(II) hydroxide because the Cu^{2+} ions have been *sequestered*.

There is a thermodynamic reason for the stability of metal chelates. Consider the equation for the reaction of $EDTA^{4-}$ given above. The left-hand side of the equation has *two* particles, $[M(H_2O)_6]^{2+}$ and $EDTA^{4-}$, whereas the right-hand side of this equation has *seven* particles, $[M(EDTA)]^{2-}$ and six H_2O molecules. This considerable increase in randomness results in an increase in *entropy* (see this book, section 3.1.8.2).

Recall that a chemical reaction becomes *feasible* if the change in free energy, $\Delta G^{\ominus}$, is negative or zero. The equation:

$$\Delta G^{\ominus} = \Delta H^{\ominus} - T\Delta S^{\ominus}$$

shows that there will *always* be a negative free-energy change if the enthalpy change, $\Delta H^{\ominus}$, is *negative* and the entropy change, $\Delta S^{\ominus}$, is *positive*. Even when $\Delta H^{\ominus}$ is *positive* (but not very much so), it will readily be outweighed by the $T\Delta S^{\ominus}$ term if the entropy change is *large and positive*.

Notes

Use of this **sequestering ability** is made in water softening, where the deposition of calcium carbonate in pipes can be prevented.

In the formation of a chelate, $\Delta S^{\ominus}$ is indeed *large and positive* (two particles form seven particles). $\Delta H^{\ominus}$ for such reactions is usually quite small because the bonds formed are the same in number and rather similar in strength to the bonds broken. Even if $\Delta H^{\ominus}$ were to be slightly positive, this term would be outweighed heavily by $-T\Delta S^{\ominus}$. Consequently, $\Delta G^{\ominus}$ is *always very negative* and the reaction is *always feasible*. Reactions of this kind are sometimes described as being **entropy driven**, because the enthalpy term can usually be ignored.

Notes

The increased entropy leads to a large and positive $T\Delta S^{\ominus}$ term which, because it appears in the equation with a negative sign, contributes significantly to ensuring that $\Delta G^{\ominus}$ is very negative.

3.2.5.3 Shapes of complex ions

The most common shape of complex ions is the **octahedral** shape formed, for example, when water molecules co-ordinate to a copper(II) ion. All the transition metals of the first-row transition series form octahedral hexaaqua ions with water. Ammonia is another ligand with no charge; it is similar in size to water and readily forms octahedral complexes with these metals.

The next most common shape encountered in complexes is **tetrahedral**. If the ligand is large and negatively charged (so that there is inter-ligand repulsion as well as bond-pair repulsion in the complex), then it may not be possible to fit six ligands around the central metal ion in a stable complex, so the tetrahedral arrangement is preferred. Ligands are further apart in a tetrahedral complex than they are in an octahedral complex. Chloride ions, bromide ions and iodide ions are typical of large anions which form tetrahedral complexes.

Notes

Fluoride ions do form octahedral complexes, e.g. $[AlF_6]^{3-}$ and $[FeF_6]^{3-}$, because the F^- ion is smaller than the Cl^- ion.

Less common shapes include the **square-planar** shape in, for example, cisplatin (see below) and the **linear** shape commonly seen in silver complexes, for example, $[Ag(NH_3)_2]^+$, used in **Tollens' reagent** to distinguish between aldehydes and ketones (see this book, section 3.2.5.5)

The arrangements of the bonds in the different shapes of complexes are shown in Fig 22.

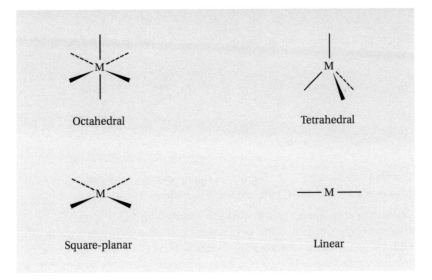

Fig 22
The shapes of complex ions

Isomerism in complexes

Octahedral complexes can display **E–Z stereoisomerism** (geometrical or *cis–trans* isomerism) with monodentate ligands, as shown in Fig 23. Note that in the *cis* isomer the two chloride ligands are at 90° to one another, whereas in the *trans* isomer they are at 180°.

Fig 23
Cis and *trans* isomers of $[Co(NH_3)_4Cl_2]$

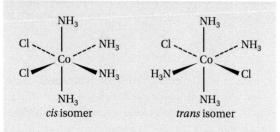

cis isomer   *trans* isomer

With bidentate ligands, octahedral complexes can show **optical isomerism**, for example as shown in Fig 24. Note that the full structure of the bidentate ligand $(H_2NCH_2CH_2NH_2)$ is not shown, only the nitrogen atoms forming co-ordinate bonds to the central Co^{2+} ion. The overall charges on the complexes are not shown either. It can be seen that the two optical isomers are non-superimposable mirror images of one another.

Fig 24
Optical isomers of $[Co(H_2NCH_2CH_2NH_2)_3]$

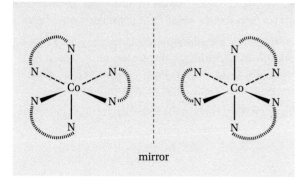

mirror

Square planar complexes can also display *cis–trans* isomerism, for example, platin (Fig 25).

Fig 25
Cis and *trans* isomers of platin

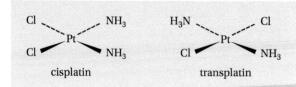

cisplatin   transplatin

The platin isomers are four-co-ordinate square-planar complexes of platinum(II), with two Cl^- ions and two NH_3 molecules as ligands.

The isomer **cisplatin** is a well-known anti-cancer drug: its use is in chemotherapy. Cisplatin works by undergoing a ligand exchange reaction with one of the nitrogen atoms of a guanine base on DNA. The drug is usually injected in the form of $(NH_3)_2PtCl_2$. This neutral complex, after passing through the cell membrane of a cancer cell, reacts with water to form the complex ion $[(NH_3)_2PtCl(H_2O)]^+$. Even though the guanine is still bonded to the deoxyribose in a DNA strand, its nitrogen 7 forms a co-ordinate bond to the platinum, replacing water (see Fig 26). Another guanine base on the same strand of DNA or on the complementary strand then bonds to the platinum by replacing the chloride ligand. If the same strand of DNA is bonded twice to the platinum, the strand becomes kinked and will not replicate. If the two DNA strands are bonded to the platinum, this prevents the strands from separating and replication is prevented in this case as well.

Fig 26
Guanine base attached to DNA. Note that the guanine structure is given on the AQA data sheet.

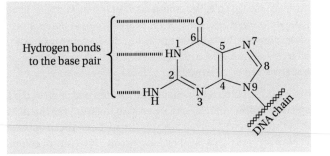

3.2.5.4 Formation of coloured ions

Many of the coloured solutions which are seen around the laboratory owe their colour to the presence of a transition metal. Indeed, along with other evidence, the colour of a solution can be used to identify the presence of a particular transition metal.

Origin of colour

Colour arises when a molecule or an ion absorbs some of the wavelengths of visible light and enters a higher or *excited* energy state. If only a part of the visible spectrum is absorbed, then the eye detects those wavelengths which remain and a colour is seen.

Energy is absorbed when electrons are promoted from the **ground state** to a higher energy level, as shown in Fig 27. The energy difference between the two levels involved is given by:

$$E_2 - E_1 = \Delta E = h\nu = \frac{hc}{\lambda}$$

where h is **Planck's constant**, ν is the frequency of the absorbed radiation, λ is the wavelength of the absorbed radiation and c is the speed of light.

Essential Notes

$h = 6.63 \times 10^{-34}$ J s

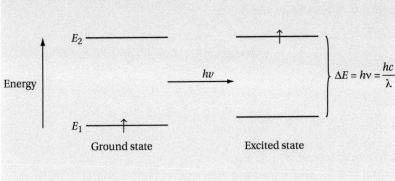

Fig 27
Absorption of light giving rise to colour. The symbol ↑ represents an electron.

Essential Notes

If both levels involved are levels of d orbitals, the change is called a *d–d transition* and the associated colour is not very intense. If, however, the transition involves an electron moving from a ligand orbital to a metal (d) orbital (or vice-versa), the transition is called *charge-transfer* and an intense colour results.

The energy difference, ΔE, between the two d-electron energy levels in a complex varies according to the oxidation state of the transition metal, the co-ordination number and the ligands.

Essential Notes

The **Beer–Lambert law:** $A = \varepsilon cl$ links absorbance (A) to concentration (c), cell path-length (l) and molar absorption coefficient (ε).

Notes

The colorimeter is a relatively cheap instrument that is used in industry and research to measure the concentration of coloured species in a rapid and non-destructive manner. Colorimeters can be operated remotely and the output fed to a computer for rapid analysis or data storage.

Notes

See the hydrolysis reactions later for an explanation of the brown colour seen in iron(III) solutions.

Notes

In some reactions, the colour change arises as a result of all three changes, e.g.

$[Mn(H_2O)_6]^{2+} \rightarrow [MnO_4]^-$
Mn(II)   Mn(VII)
very pale pink   purple
octahedral   tetrahedral

Visible spectrophotometry

The intensity of colours shown by a solution of a transition-metal ion can be used to determine the concentration of that ion; this is done using a spectrophotometer.

In a **visible spectrophotometer**, light of increasing (or decreasing) wavelength is passed through the sample; the emergent light is received in a detector and recorded. The *amount of light absorbed is proportional to the concentration of the absorbing species* in the solution under test. By measuring how much light is absorbed by a solution at a particular wavelength, the concentration of the absorbing species can be determined.

So sensitive are these colours to a change in the concentration of the ion, that a very simple spectrophotometer can be used to measure the intensity of absorption at the particular wavelength absorbed. Such spectrophotometers are called **colorimeters** (because they measure the intensity of colours).

Factors leading to colour change

When a colour change occurs in the reaction of a transition-metal ion, there is a change in at least one of the following factors:

- oxidation state
- co-ordination number
- ligand.

Often there is a change in more than one of these factors.

Change of oxidation state

An example of a change in oxidation state being responsible for the colour change is:

$$[Fe^{II}(H_2O)_6]^{2+} \quad \rightarrow \quad [Fe^{III}(H_2O)_6]^{3+}$$
green   very pale violet

Change of co-ordination number

A change in co-ordination number (i.e. the number of nearest neighbours) is most usually achieved by changing ligands. Examples of colour changes arising from this type of change are:

$$[Cu(H_2O)_6]^{2+} \quad \rightarrow \quad [CuCl_4]^{2-}$$
blue   yellow-green
octahedral   tetrahedral

$$[Co(H_2O)_6]^{2+} \quad \rightarrow \quad [CoCl_4]^{2-}$$
pink   blue
octahedral   tetrahedral

Change of ligand

Colour changes arising from a change of ligand only are shown by the reactions:

$$[Co(H_2O)_6]^{2+} \quad \rightarrow \quad [Co(NH_3)_6]^{2+}$$
pink   pale brown
octahedral   octahedral

$$[Cu(H_2O)_6]^{2+} \quad \rightarrow \quad [Cu(NH_3)_4(H_2O)_2]^{2+}$$

blue   deep blue

octahedral   octahedral

3.2.5.5 Variable oxidation states

A characteristic property of the transition elements is their ability to change oxidation state in chemical reactions.

In an oxidation–reduction or 'redox' reaction, the transition-metal ion is oxidised or reduced, i.e. it changes its oxidation state. Because transition-metal ions show variable oxidation states, many redox reactions occur.

Metallic zinc is a good reducing agent and reduces aqueous transition-metal ions to low oxidation states (see *Collins Student Support Materials: AS/A-Level year 1 – Inorganic and Relevant Physical Chemistry*, section 3.1.7). Zinc reacts with mineral acids to form zinc(II) ions and releases two electrons for the reduction:

$$Zn \rightarrow Zn^{2+} + 2e^-$$

Reductions with zinc proceed through the oxidation states of the metal in sequence, so that each different colour of the metal ion in every oxidation state can often be seen. A good example of this behaviour is seen in the reduction of vanadium(V).

Oxidation states of vanadium

When white ammonium vanadate(V), NH_4VO_3, is added to dilute hydrochloric acid, the orange colour of the dioxovanadium(V) ion, VO_2^+, is seen

$$VO_3^- + 2H^+ \rightarrow VO_2^+ + H_2O$$

white   orange

The addition of granulated zinc to this mixture results in colour changes which occur over several minutes, as the reduction proceeds. The oxidation states, and colours of the ions seen, are:

V(V)	$VO_2^+(aq)$		yellow
↓	↓		↓
V(IV)	$VO^{2+}(aq)$	i.e. $[VO(H_2O)_5]^{2+}$	blue
↓	↓		↓
V(III)	$V^{3+}(aq)$	i.e. $[VCl_2(H_2O)_4]^+$	green
↓	↓		↓
V(II)	$V^{2+}(aq)$	i.e. $[V(H_2O)_6]^{2+}$	violet

If sulfuric acid is used instead of hydrochloric acid in this reduction, the vanadium(III) species seen will have the dull grey-blue colour of the hexaaqua ion $[V(H_2O)_6]^{3+}$. Since vanadium(II) ions are oxidised by air, it may be necessary to carry out this reduction in a lightly stoppered flask.

Effect of pH and ligand change on redox potential

The relative stability of different oxidation states of a transition metal changes when the pH is altered, for example, iron(II) and iron(III). Consider the following $E^\ominus$ values:

In neutral solution: $\quad Fe^{3+}(aq) + e^- \rightarrow Fe^{2+}(aq)$ $\hspace{3cm}$ $E^\ominus = +0.77$ V

In alkaline solution: $\quad Fe(OH)_3(s) + e^- \rightarrow Fe(OH)_2(s) + OH^-(aq)$ $\quad E^\ominus = -0.56$ V

$\hspace{3.2cm} O_2(g) + 2H_2O(l) + 4e^- \rightarrow 4OH^-(aq)$ $\hspace{1.2cm}$ $E^\ominus = +0.40$ V

Since a more negative/less positive $E^\ominus$ indicates greater reducing power of the species on the right-hand side, this shows that at higher pH, iron(II), in the form of $Fe(OH)_2$, is a stronger reducing agent than $Fe^{2+}(aq)$ and so is more readily oxidised to iron(III).

Furthermore, since $E^\ominus(O_2/OH^-)$ is more positive/less negative than $E^\ominus(Fe(OH)_3/Fe(OH)_2)$, oxygen can oxidise $Fe(OH)_2$ to $Fe(OH)_3$. This is observed experimentally in the rapid darkening of the green precipitate $Fe(OH)_2$ as it changes to the brown $Fe(OH)_3$ in the presence of air.

At lower pH, i.e. in the presence of acid, the oxidation of iron(II) to iron(III) is more difficult still and so solutions of iron(II) salts are made up in dilute sulfuric acid to stabilise the iron(II).

Similarly, when transition metals are complexed, redox potentials change. Again, iron(II) and iron(III) can be used as illustration. Consider the following $E^\ominus$ values:

$$Fe^{3+}(aq) + e^- \rightarrow Fe^{2+}(aq) \hspace{2cm} E^\ominus = +0.77 \text{ V}$$
$$Fe(CN)_6^{3-}(aq) + e^- \rightarrow Fe(CN)_6^{4-}(aq) \hspace{1cm} E^\ominus = +0.36 \text{ V}$$
$$I_2(aq) + 2e^- \rightarrow 2I^-(aq) \hspace{2.3cm} E^\ominus = +0.54 \text{ V}$$

Again, since a more negative/less positive $E^\ominus$ indicates greater reducing power of the species on the right-hand side, this shows that in complexed form, iron(II), as $Fe(CN)_6^{4-}$, is a weaker reducing agent than $Fe^{2+}(aq)$ and so $Fe(CN)_6^{4-}$ is less readily oxidised to iron(II).

Furthermore, since $E^\ominus(I_2/I^-)$ is less positive than $E^\ominus(Fe^{3+}/Fe^{2+})$, iodide ions can reduce Fe^{3+} to Fe^{2+}. This is observed experimentally when the brown colour of iodine is seen on addition of potassium iodide to a solution of any iron(III) salt.

However, since $E^\ominus(I_2/I^-)$ is more positive than $E^\ominus(Fe(CN)_6^{3-}/Fe(CN)_6^{4-})$, iodide ions cannot reduce $Fe(CN)_6^{3-}$ to $Fe(CN)_6^{4-}$. Hence there is no brown colour of iodine when potassium iodide is added to a solution containing $Fe(CN)_6^{3-}$.

Oxidation states of silver

In section 3.2.5.3 of this book it was noted that silver(I) ions usually form linear complexes. The complex ion $[Ag(NH_3)_2]^+$ is formed when silver chloride or silver oxide dissolves in aqueous ammonia to form the colourless solution known as Tollens' reagent. This reagent is used to distinguish aldehydes from ketones, since the silver(I) complex is reduced to silver(0) by aldehydes. Aldehydes give a silver mirror on the inside of the test tube, ketones do not (see *Collins Student Support Materials: AS/A-Level year 1 – Organic and Relevant Physical Chemistry*, section 3.3.5.2):

$$RCHO + 2[Ag(NH_3)_2]^+ + 3OH^- \rightarrow RCOO^- + 2Ag + 4NH_3 + 2H_2O$$

Oxidation states of manganese

Manganese shows its maximum oxidation state of +7 in the manganate(VII), MnO_4^-, ion. This ion is a powerful oxidising agent which is reduced directly to the manganese +2 state by zinc, by iron, or even by iron(II) ions in acid solution.

The $[Mn(H_2O)_6]^{2+}$ ion is a very pale pink colour and appears colourless in aqueous solution so that, when manganate(VII) ions are reduced to Mn(II) ions, the colour change seen is from purple to colourless:

Mn(VII)	MnO_4^-	purple
$\downarrow$	$\downarrow$	$\downarrow$
Mn(II)	$[Mn(H_2O)_6]^{2+}$	colourless

Redox titrations

Redox reactions can be used in quantitative volumetric analysis. The end-point, when the reaction is complete, can be determined in a variety of ways, including the use of indicators or electrical methods. The use in **redox titrations** of one common quantitative oxidising agent is discussed below.

Potassium manganate(VII) titrations

When potassium manganate(VII) is used as an oxidising agent in acidic solution, no additional indicator is required. Because the reduction product, $[Mn(H_2O)_6]^{2+}$, is essentially colourless, no colour is seen until an excess of potassium manganate(VII) has been added, at which point the presence of even a slight excess of the dark purple manganate(VII) ion is readily seen and is used to indicate complete reaction. Knowing the initial and final species present, MnO_4^- and Mn^{2+}, the following half-equation can be constructed:

$$MnO_4^- + 8H^+ + 5e^- \rightarrow Mn^{2+} + 4H_2O$$

The presence of an excess of acid for this reaction is essential and the choice of the actual acid used is important. The acid must:

- be strong, because a high concentration of hydrogen ions is needed

- not be an oxidising agent, because an oxidising agent might react with the reducing agent being estimated

- not be a reducing agent, because reducing agents will be oxidised by the manganate(VII) ions.

Notes

If insufficient acid is present, a brown colouration of insoluble MnO_2 is seen.

A review of common laboratory acids enables the correct choice of acid to be made:

- it *cannot* be hydrochloric acid, as this can be oxidised to chlorine
- it *cannot* be nitric acid, as this is an oxidising agent
- it *cannot* be *concentrated* sulfuric acid, as this too is an oxidising agent
- it *cannot* be ethanoic acid, as this is a *weak* acid which produces too low a concentration of H^+ ions.

Dilute sulfuric acid is, therefore, the main choice among common laboratory acids.

The usual procedure is to add a standard solution of potassium manganate(VII) from a burette to a conical flask containing a solution of the reducing agent which has been acidified by the addition of an excess of dilute sulfuric acid. As mentioned above, no additional indicator is required for this titration, the end-point occurring when a permanent pink colour is observed.

Consider, for example, the use of potassium manganate(VII) to estimate the concentration of iron(II) ions in solution. Knowing that the oxidation of iron(II) to iron(III) is a one-electron change and that the conversion of manganese(VII) into manganese(II) is a five-electron change, it is easy to deduce that the complete reduction of one mole of manganese(VII) to manganese(II) will require five moles of iron(II). The two half-equations and the **overall equation** for these processes are:

$$MnO_4^- + 8H^+ + 5e^- \rightarrow Mn^{2+} + 4H_2O$$
$$5Fe^{2+} \rightarrow 5Fe^{3+} + 5e^-$$
$$\overline{5Fe^{2+} + MnO_4^- + 8H^+ \rightarrow 5Fe^{3+} + Mn^{2+} + 4H_2O}$$

Similarly, potassium manganate(VII) can be used to estimate the concentration of ethanedioate ions in solution. The oxidation of ethanedioate to carbon dioxide is a two-electron change. The two half-equations can be combined to produce the overall equation, as follows:

$$MnO_4^- + 8H^+ + 5e^- \rightarrow Mn^{2+} + 4H_2O \qquad \times 2$$
$$C_2O_4^{2-} \rightarrow 2CO_2 + 2e^- \qquad \times 5$$
$$\overline{2MnO_4^- + 5C_2O_4^{2-} + 16H^+ \rightarrow 2Mn^{2+} + 10CO_2 + 8H_2O}$$

Example

3.00 g of a lawn sand containing an iron(II) salt were shaken with dilute sulfuric acid. The resulting solution required 25.00 cm^3 of a 0.0200 mol dm^{-3} solution of $KMnO_4$ to oxidise all the Fe^{2+} ions in the solution to Fe^{3+} ions. Calculate the percentage by mass of Fe^{2+} ions in the lawn sand.

Method

Calculate the amount, in moles, of manganate(VII) that have reacted, and then use the stoichiometric equation to relate this number to the number of amount of Fe(II) that have been oxidised. Use the relative atomic mass of iron to convert moles into mass and then express the mass as a percentage of the original sample.

Answer

$$\text{moles KMnO}_4 = \frac{0.0200 \times 25.00}{1000} = 5.00 \times 10^{-4} \text{ mol}$$

There are *five* moles of Fe^{2+} for every mole of MnO_4^- used, so

moles $Fe^{2+} = 5 \times 5.00 \times 10^{-4} = 2.50 \times 10^{-3}$ mol

Multiplying the amount of moles by the mass of one mole of Fe^{2+} gives the mass of Fe^{2+} present in 3.00 g of lawn sand:

mass of Fe^{2+} present $= 55.8 \times 2.5 \times 10^{-3} = 0.140$ g

Hence, the percentage by mass of Fe^{2+} ions present in the sample is

$$\frac{0.140 \times 100}{3.00} = 4.65\%$$

Comment

Note that it is not necessary to filter off the sand in this analysis, because it is inert and does not interfere with the titration.

Note also that *rounding off* the mass of Fe^{2+} to 3 significant figures *before* calculating the percentage gives a slightly different, and less accurate, answer.

Example

2.270 g of hydrated ethanedioic acid, $H_2C_2O_4.xH_2O$, were dissolved in water and made up to 250 cm³. 25 cm³ of this solution required 36.00 cm³ of 0.0200 mol dm^{-3} potassium manganate(VII) solution for oxidation. Calculate x.

Method

Calculate the moles of manganate(VII) that have reacted and then use the stoichiometric equation to relate this to the amount of moles of ethanedioate that have been oxidised in the titration. Scale this up by a factor of 10 to give total amount of moles of ethanedioate in the sample. Calculate the M_r of anhydrous ethanedioic acid and use it to calculate the total mass of anhydrous ethanedioic acid present. Subtract this from the initial mass of hydrated ethanedioic acid to get the mass of water in the sample. Convert the mass ratio of anhydrous ethanedioic acid to water into a mole ratio.

$$\text{moles KMnO}_4 = \frac{0.0200 \times 36.00}{1000} = 7.200 \times 10^{-4} \text{ mol}$$

There are *five* moles of ethanedioate for every two moles of MnO_4^- used, so

moles $C_2O_4^{2-}$ in 25 cm³ $= \frac{5}{2} \times 7.200 \times 10^{-4} = 1.800 \times 10^{-3}$, so

moles $C_2O_4^{2-}$ in 250 cm³ $= 10 \times 1.375 \times 10^{-3} = 1.800 \times 10^{-2}$

$M_r(H_2C_2O_4) = 2.0 + 24.0 + 64.0 = 90.0$, so

mass of anhydrous $H_2C_2O_4$ in 250 cm³ $= 1.800 \times 10^{-2} \times 90.0$

$$= 1.620 \text{ g}$$

Therefore 2.270 g $H_2C_2O_4.xH_2O$ contains 1.620 g anhydrous $H_2C_2O_4$

and $2.270 - 1.620 = 0.650$ g H_2O

$$\text{moles anhydrous } H_2C_2O_4 = \frac{1.620}{90} = 0.0180$$

$$\text{moles } H_2O \qquad \qquad = \frac{0.650}{18} = 0.0361, \text{ so}$$

mole ratio $H_2C_2O_4$:H_2O is 0.180:0.0361

i.e. 1:2

Therefore, $x = 2$

3.2.5.6 Catalysts

A **catalyst** (see *Collins Student Support Materials: AS/A-Level year 1 – Organic and Relevant Physical Chemistry,* section 3.1.5.5) makes reactions go faster, i.e. it increases the rate of attainment of equilibrium. Catalysts are classified according to whether they act in a different phase from, or in the same phase as, the reactants.

Notes

The phase may be solid or liquid or gas.

Definition

A catalyst which acts in a **different phase** from the reactants is called a **heterogeneous** catalyst.

A catalyst which acts in the **same phase** as the reactants is called a **homogeneous** catalyst.

Because the success of an industrial process depends upon making the maximum amount of product in the shortest possible time, catalysis is at the heart of the chemical manufacturing industry. Most of the catalysts used in industry are transition metals or their compounds.

Heterogeneous catalysis

Heterogeneous catalysis depends on at least one reactant being adsorbed onto active sites on the catalyst surface (usually by weak chemical bonds), and being modified into a state which makes reaction more likely.

Sometimes **adsorption** onto the surface can weaken reactant bonds, or induce a reactant molecule to break up into very reactive fragments; but equally, adsorption may be responsible for holding a reactant molecule on the surface in exactly the right configuration to make reaction easier.

In simple gas reactions such as:

$$A(g) + B(g) \rightarrow \text{products}$$

catalysis leading to an increase in reaction rate can come about if:

- **A** is adsorbed on the surface, becomes more accessible to collisions with **B**, and so reacts more readily than by random collisions in the gas phase. *Collision frequency increases.*

- **A** is held on the surface of the catalyst in a particularly favourable and highly reactive configuration. Unadsorbed **B** collides with reactive **A**. *Activation energy decreases.*

- **A** is adsorbed onto the surface and undergoes internal bond-breaking or rearrangement. Reactive fragments can easily react. *Activation energy decreases.*

Making catalysts more effective

Heterogeneous catalysts act through surface properties. Increasing the surface area available for reaction is always worthwhile. A wafer-thin metal foil is much more active catalytically than is the same mass of metal in a solid lump. Expensive catalysts are frequently spread very thinly or impregnated onto an **inert support medium** in order to maximise the **surface-to-mass ratio**. Solids which have large surface areas in their own right (e.g. silica) are useful as support media, especially if ground into very fine powders to maximise the surface area.

The practice of increasing surface area leads to more effective catalysis and also to the much more effective use of high-cost catalytic agents. *Loss of catalyst from a support can greatly increase the inherent cost of a catalytic process, not least by decreasing the overall efficiency of the reaction.*

Some catalysts can combine surface and support functions. The support itself may show catalytic activity and such **mixed catalysts** can often perform dual functions.

The most useful catalysts, such as Pt and Rh, are also among the most expensive. **Catalytic converters** in cars are designed to give a *high surface-area* of active catalyst and use rhodium spread thinly on a cheap ceramic support. Modern catalysts in car exhaust converters reduce noxious emissions to a tolerable level.

Manufacture of methanol

Chromium(III) oxide, Cr_2O_3, in a mixture with zinc and/or copper(II) oxides, is used as a catalyst in the synthesis of methanol from carbon monoxide and hydrogen:

$$CO + 2H_2 \rightarrow CH_3OH$$

The Haber and Contact processes

Two large-scale industrial processes use heterogeneous catalysis by a transition metal or one of its compounds. In the production of ammonia from nitrogen and hydrogen by the **Haber process**, iron is the catalyst. In the manufacture of sulfuric acid by the **Contact process**, the conversion of sulfur dioxide into sulfur trioxide is catalysed by vanadium(V) oxide. The sulfur trioxide reacts with water to give sulfuric acid. These two processes are summarised below.

Haber process : $\quad N_2(g) + 3H_2(g) \xrightleftharpoons{\text{Fe(s)}} 2NH_3(g)$

Contact process : $\quad 2SO_2(g) + O_2(g) \xrightleftharpoons{\text{V}_2\text{O}_5\text{(s)}} 2SO_3(g)$

The importance of the variable oxidation state of vanadium in the Contact process can be seen by considering how the catalyst works. Sulfur dioxide is oxidised to sulfur trioxide by vanadium(V) oxide:

$$SO_2 + V_2O_5 \rightarrow SO_3 + V_2O_4$$

The vanadium(V) has thus been reduced to vanadium(IV), which is then oxidised back to vanadium(V) oxide by oxygen:

$$2V_2O_4 + O_2 \rightarrow 2V_2O_5$$

Thus, the vanadium(V) oxide is unchanged at the end of the reaction and acts as a catalyst by use of the two oxidation states of vanadium, +5 and +4. The presence of the vanadium(V) oxide enables the reaction to proceed by a different route with a lower activation energy.

Notes

Anti-knock additives such as tetraethyl-lead are no longer used in the UK.

Notes

Catalyst poisons need to be removed from feedstock gases in the Haber process. The iron catalyst is poisoned by CO, CO_2 and water vapour, and also by sulfur compounds which are contaminants of the natural gas used to produce hydrogen. This increases the operating costs.

Notes

The two steps of the catalysed reaction can occur in either order, so that iron(III) ions can also be used as the catalyst.

Catalyst poisons

Surface-active catalysts are especially prone to *poisoning*. For example, catalytic converters for car exhausts are readily poisoned by lead in *anti-knock* additives.

Poisoning occurs when unwanted contaminants, or waste products, become adsorbed too strongly onto the surface. Transition-metal catalysts, for example, are especially prone to poisoning by sulfur compounds, which form inactive surface sulfides. The effect is cumulative. The sulfur content of feedstock gases in metal-catalysed reactions must be reduced to a minimum.

Homogeneous catalysis

Reactions which are catalysed homogeneously by a transition-metal ion usually involve a change in oxidation state of the transition metal during catalysis. Such reactions proceed through the formation of an intermediate species, which can form because the metal has variable oxidation states.

The oxidation of iodide ions by peroxodisulfate(VI) ions catalysed by iron ions

The reaction:

$$S_2O_8^{2-} + 2I^- \rightarrow I_2 + 2SO_4^{2-}$$

is very slow even though it is energetically very favourable. Both ions are negatively charged and so are unlikely to make effective collisions with each other. The reaction occurs rapidly, however, when iron(II) ions are added. These cationic species can make effective collisions with anions. Iron(II) ions are oxidised by peroxodisulfate(VI) ions; the iron(II) becomes iron(III) which, in turn, can rapidly oxidise iodide ions to iodine:

$$2Fe^{2+} + S_2O_8^{2-} \rightarrow 2Fe^{3+} + 2SO_4^{2-}$$
$$\underline{2Fe^{3+} + 2I^- \rightarrow 2Fe^{2+} + I_2}$$
$$S_2O_8^{2-} + 2I^- \rightarrow I_2 + 2SO_4^{2-}$$

The variable oxidation state of iron enables the reaction to proceed by an alternative route with a lower activation energy.

Electrode potentials show how both Fe^{2+} and Fe^{3+} can catalyse the peroxodisulfate/iodide reaction:

$$S_2O_8^{2-}(aq) + 2e^- \rightarrow 2SO_4^{2-}(aq) \qquad E^\ominus = +2.01 \text{ V}$$
$$Fe^{3+}(aq) + e^- \rightarrow Fe^{2+}(aq) \qquad E^\ominus = +0.77 \text{ V}$$
$$I_2(aq) + 2e^- \rightarrow 2I^-(aq) \qquad E^\ominus = +0.54 \text{ V}$$

Catalysis by Fe^{3+}

Step 1: $E^\ominus(Fe^{3+}/Fe^{2+})$ is more positive than $E^\ominus(I_2/I^-)$, and so Fe^{3+} can oxidise I^- to I_2, with Fe^{3+} undergoing reduction to Fe^{2+}.

Step 2: Since $E^\ominus(S_2O_8^{2-}/SO_4^{2-})$ is more positive than $E^\ominus(Fe^{3+}/Fe^{2+})$, $S_2O_8^{2-}$ can oxidise the resulting Fe^{2+} back to Fe^{3+}, with $S_2O_8^{2-}$ undergoing reduction to SO_4^{2-}.

Catalysis by Fe^{2+}

The two steps are the same, but in reverse order. Fe^{2+} is oxidised by $S_2O_8^{2-}$, which undergoes reduction to SO_4^{2-}. The resulting Fe^{3+} then oxidises the I^- to I_2. The Fe^{2+} is oxidised to Fe^{3+}, thus reforming the catalyst.

Autocatalysis by manganese ions in the reaction between ethanedioate ions and manganate(VII) ions

When a solution of potassium manganate(VII), from a burette, is run into a hot, acidified solution containing ethanedioate ions, the purple colour is not decolourised immediately in the early stages of the titration. Once the initial purple colour produced by the addition of the first few drops of manganate(VII) solution has been discharged, however, further addition of the oxidant leads to immediate decolourisation. This rapid decolourisation continues until the end-point is reached.

The overall reaction is:

$$2MnO_4^- + 16H^+ + 5C_2O_4^{2-} \rightarrow 2Mn^{2+} + 10CO_2 + 8H_2O$$

The reaction between MnO_4^- ions and $C_2O_4^{2-}$ ions is slow; as in the previous example, both ions are negatively charged and so are less likely to make effective collisions. However, the reaction is catalysed by Mn^{2+} ions. Once some Mn^{2+} ions have been formed in the early stages of the titration, they can react with MnO_4^- ions to form a few Mn^{3+} ions:

$$4Mn^{2+} + MnO_4^- + 8H^+ \rightarrow 5Mn^{3+} + 4H_2O$$

These Mn^{3+} ions can then react with $C_2O_4^{2-}$ ions to liberate CO_2 and re-form Mn^{2+} ions:

$$2Mn^{3+} + C_2O_4^{2-} \rightarrow 2CO_2 + 2Mn^{2+}$$

So, until some Mn^{2+} ions have been formed, the manganate(VII) solution is decolourised only slowly. The reaction then speeds up as the Mn^{2+} formed acts as a catalyst, eventually slowing down as the MnO_4^- is used up. Such catalysis by a product of the same reaction is called **autocatalysis**.

Fig 28 shows how the reagent concentration changes over time.

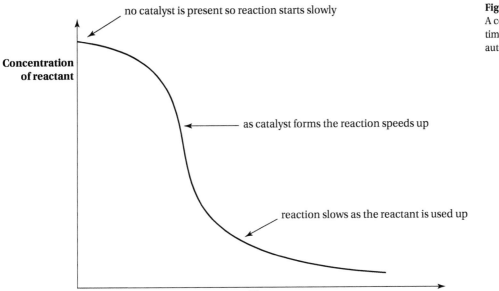

no catalyst is present so reaction starts slowly

Concentration of reactant

as catalyst forms the reaction speeds up

reaction slows as the reactant is used up

Time

Fig 28
A concentration-time graph for an autocatalysed reaction

3.2.6 Reactions of ions in aqueous solution

Metal-aqua ions

When a transition-metal ion is placed in water, one of the two lone pairs of electrons on each of six water molecules is used to form a co-ordinate bond with the metal ion. The result is a **co-ordination compound** or **complex**. This reaction and its associated colour change was illustrated earlier in this book (section 3.2.5.1), using copper(II) sulphate.

A still more striking colour change occurs when anhydrous cobalt(II) chloride is added to water. The anhydrous salt is blue, whereas the hexaaquacobalt(II) cation is pink.

The colours of solutions of transition-metal salts in water are largely due to the colours of the metal-aqua ions, for example:

$[Fe(H_2O)_6]^{2+}$ green

$[Co(H_2O)_6]^{2+}$ pink

$[Cu(H_2O)_6]^{2+}$ blue

Any copper(II) salt in aqueous solution will be present as the hexaaqua ion and the solution will be blue. Anhydrous copper(II) salts are not blue: $CuSO_4$ is white, $CuCl_2$ is yellow, $CuBr_2$ is black, and CuO is also black, yet they all give blue solutions when dissolved in water or dissolved in dilute acids.

So strong is the bonding between the water molecules and the metal ions, that the hexaaqua ion remains intact when solutions containing this ion are evaporated. Thus, solid transition-metal salts in the laboratory are often not simple salts but contain a complex, the metal-hexaaqua ion. Examples of 'salts' containing the hexaaqua ion in the solid state include:

$FeSO_4.7H_2O$	green	contains $[Fe(H_2O)_6]^{2+}$
$CoCl_2.6H_2O$	pink	contains $[Co(H_2O)_6]^{2+}$
$Fe(NO_3)_3.9H_2O$	pale violet	contains $[Fe(H_2O)_6]^{3+}$

Metal(III)-aqua ions

Whilst the reactions of metal(II) salts with water do not appear to be violent, enough energy is given out on hydration to overcome the lattice energy of the solid so that the solid dissolves; reaction with water breaks down the crystal lattice. When the charge on the metal ion is increased to 3+, however, the hydration reaction is often very exothermic. Thus, aluminium chloride dissolves in water violently with evolution of heat and fumes:

$$AlCl_3(s) \xrightarrow{\text{excess } H_2O(l)} [Al(H_2O)_6]^{3+}(aq) + 3Cl^-(aq)$$

Note that there is no precipitate of a metal hydroxide in this reaction; in general, when simply added to water, metal(III) chlorides do not hydrolyse to form metal(III) hydroxides.

Metal ions exist in aqueous solution as 6-water or hexaaqua ions. The hexaaqua ion contains a metal ion surrounded octahedrally by six oxygen atoms from six water molecules. When metal ions react in aqueous solution, it is essential to

Notes

Note that the anion plays no part in the final colour seen. If green cobalt(II) bromide is added to water, the same pink hexaaquacobalt(II) cation is formed as with cobalt(II) chloride. The bromide and chloride ions in these reactions become aquated, forming colourless Br^-(aq) ions and Cl^-(aq) ions, respectively.

remember these six water molecules because they can take part in the chemical reactions of metal ions.

When chemical reactions occur, bonds are frequently broken and, more often than not, new bonds are formed. The driving force for reaction is often bond formation. When considering reactions in water, the questions to ask are, 'which bonds can be broken?' and 'which bonds can be formed?'

The bonding in a hexaaqua ion, $[M(H_2O)_6]^{n+}$, is shown in Fig 29.

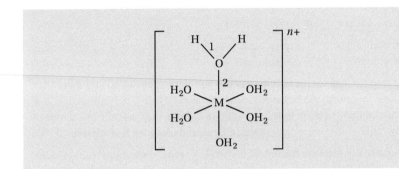

Fig 29
The hexaaqua ion $[M(H_2O)_6]^{n+}$ contains only *two* types of bond:

O—H bonds, covalent bonds in water molecules co-ordinated to the metal ion; one of these bonds is labelled 1.

M—O bonds, co-ordinate bonds between water ligands and the metal ion; one of these bonds is labelled 2.

In reactions of metal ions, the only process which occurs without breaking either of these two types of bond is a redox reaction, i.e. oxidation or reduction, which involves only the loss or gain of an electron, for example:

$$[M(H_2O)_6]^{2+} \rightarrow [M(H_2O)_6]^{3+} + e^-$$

If the O—H bond in a co-ordinated water molecule is broken, the reaction is called an **acidity reaction** or a **hydrolysis reaction**; if the M—O bond is broken (and a new ligand is attached to the metal) the reaction is called a **substitution reaction**. Ligand substitution reactions have been considered earlier (pages 76–83). The acidity or hydrolysis reaction is now considered in detail.

Acidity or hydrolysis reaction

When a metal-aqua ion is placed in water, an equilibrium is set up:

$$[M(H_2O)_6]^{2+} + H_2O \rightleftharpoons [M(H_2O)_5(OH)]^+ + H_3O^+$$

which is called the *hydrolysis* reaction, because a water molecule has been split into OH⁻ and H⁺. It is also called the *acidity* reaction, because hydrolysis leads to the formation of H_3O^+ ions.

For a metal(II) ion, as above, the equilibrium lies very much to the left-hand side, so that metal(II) ions are only slightly acidic in water. The degree of acidity is measured by the magnitude of the equilibrium constant for the reaction (see *Collins Student Support Materials: AS/A-Level year 1 – Inorganic and Relevant Physical Chemistry*, section 3.1.6.2). For the above equilibrium, the **acid dissociation constant** (or **acidity constant**) is defined as:

$$K_a = \frac{[[M(H_2O)_5(OH)]^+][H_3O^+]}{[[M(H_2O)_6]^{2+}]}$$

Notes

Hydrolysis is a reaction of water in which an O–H bond is broken. A reaction of water in which the water molecule remains intact is called hydration.

Essential Notes

In this relationship, the outer square brackets indicate the concentrations of the different species.

For metal(II) ions, K_a varies between 10^{-6} and 10^{-11} and $\mathbf{pK_a}$ varies between 6 and 11.

For metal(III) ions, however, the equilibrium:

$$[M(H_2O)_6]^{3+} + H_2O \rightleftharpoons [M(H_2O)_5(OH)]^{2+} + H_3O^+$$

lies further to the right and, for this equilibrium, K_a varies between 10^{-2} and 10^{-5}, so that pK_a takes values between 2 and 5.

The acidity of transition-metal ions at usual concentrations can be summarised as follows:

$[M(H_2O)_6]^{2+}$	$[M(H_2O)_6]^{3+}$
very weak acids	weak acids
pH around 6	pH around 3

The principal species present in aqueous solutions of metal(II) and metal(III) salts are, therefore, the hexaaqua ions. For metal(II) ions, only about one in a hundred thousand aqua ions has undergone the hydrolysis reaction, whereas for metal(III) ions, more than one in a thousand has been hydrolysed.

Because of the extreme hydrolysis of metal(IV) ions, the ion $[M(H_2O)_6]^{4+}$ does not usually exist in aqueous solution; it hydrolyses to form $M(OH)_4$.

Two factors are involved in deciding the acidity of metal ions; these are:

- the charge on the metal ion   *acidity increases with charge*

- the size of the metal ion   *acidity decreases as size increases*

Small, highly-charged cations are the strongest acids in aqueous solution. If the charge and size factors are taken together, the **charge-to-size ratio** can be used to predict the relative acidities of metal ions. This ratio reflects the **polarising power** of the metal ion. The greater the power of the metal ion to attract electron density from the oxygen atom of a co-ordinated water molecule, the weaker is the O–H bond in this molecule. As a result, it is easier to break this O–H bond and release an H^+ ion to a water molecule which is not co-ordinated.

So far, only the first stage of the hydrolysis reaction has been considered. Further hydrolysis can occur and this is especially true if a base is added to upset the equilibrium.

For metal(II) ions the complete hydrolysis is shown by the equations:

$$[M(H_2O)_6]^{2+} + H_2O \rightleftharpoons [M(H_2O)_5(OH)]^+ + H_3O^+$$

$$[M(H_2O)_5(OH)]^+ + H_2O \rightleftharpoons [M(H_2O)_4(OH)_2] + H_3O^+$$

while for metal(III) ions, the equations are:

$$[M(H_2O)_6]^{3+} + H_2O \rightleftharpoons [M(H_2O)_5(OH)]^{2+} + H_3O^+$$

$$[M(H_2O)_5(OH)]^{2+} + H_2O \rightleftharpoons [M(H_2O)_4(OH)_2]^+ + H_3O^+$$

$$[M(H_2O)_4(OH)_2]^+ + H_2O \rightleftharpoons [M(H_2O)_3(OH)_3] + H_3O^+$$

Note that the final product in all cases is the neutral metal hydroxide. This species has no overall charge and there is therefore little hydration energy available to overcome the lattice energy, so that such hydroxides always appear as precipitates.

- All transition-metal hydroxides are insoluble in water; they are only formed from aquated metal(II) and metal(III) ions when a base (e.g. NaOH or NH_3) is added.

Some reactions, which can be understood only by using these hydrolysis equations, will now be considered. If either some pale violet crystals of iron(III) nitrate nonahydrate, $Fe(NO_3)_3.9H_2O$, or some crystals of iron(III) ammonium alum, $Fe_2(SO_4)_3.(NH_4)_2SO_4.24H_2O$, are added to water, a brown solution forms. Both of these iron salts contain the very pale violet hexaaqua ion $[Fe(H_2O)_6]^{3+}$, and it is clearly this ion which is reacting with the water:

$$[Fe(H_2O)_6]^{3+} + H_2O \rightleftharpoons [Fe(H_2O)_5(OH)]^{2+} + H_3O^+$$

pale violet   brown

The hydrolysis product is brown and, since this ion is more intensely coloured than the hexaaqua ion, the colour of the solution is seen to be brown. Don't forget, however, that the above equilibrium still lies very far to the left, so that $[Fe(H_2O)_6]^{3+}$ remains the principal species present.

Applying the principles of equilibrium (see *Collins Student Support Materials: AS/A-Level year 1 – Inorganic and Relevant Physical Chemistry,* section 3.1.6.2) to this case allows prediction of the outcome when either acid or base is added to the system above.

If a little nitric acid is added to the brown solution, the brown colour disappears and an almost colourless (pale violet) solution results. Clearly, if an acid is added to the equilibrium system above, the equilibrium will respond by moving to the left to remove the excess of acid, and some $[Fe(H_2O)_5(OH)]^{2+}$ will be converted into $[Fe(H_2O)_6]^{3+}$. Similarly, if a base such as sodium hydroxide is added to the equilibrium system above, the base will react with H_3O^+ ions, removing some of them; the system will respond by producing more H_3O^+ ions and the equilibrium will move to the right. What is observed on adding acid, therefore, is a decrease of the dark brown colour. On adding base, the solution turns darker brown until eventually (when the next two stages of the hydrolytic equilibria have been moved to the right) a brown precipitate of iron(III) hydroxide forms.

Reactions of aqua ions with alkalis

When sodium hydroxide solution is added to a solution of a transition-metal salt, a precipitate of the metal hydroxide always appears.

For instance, consider what happens when sodium hydroxide solution is added to a solution of copper(II) sulfate. The species present in a solution of copper(II) sulfate are:

$[Cu(H_2O)_6]^{2+}$ (major species), $[Cu(H_2O)_5(OH)]^+$ and H_3O^+ (small amount)

When OH^- ions are added to this mixture, they attack the strongest acid present, which is H_3O^+:

$$OH^- + H_3O^+ \rightarrow 2H_2O$$

This reaction upsets the hydrolysis equilibrium:

$$[Cu(H_2O)_6]^{2+} + H_2O \rightleftharpoons [Cu(H_2O)_5(OH)]^+ + H_3O^+$$

and, when this equilibrium has moved completely over to the right-hand side, the next equilibrium is set up:

$$[Cu(H_2O)_5(OH)]^+ + H_2O \rightleftharpoons [Cu(H_2O)_4(OH)_2] + H_3O^+$$

Notes

Metal hydroxides can be written as $M(OH)_2$ and $M(OH)_3$. Whilst the hydroxides $M(OH)_2$ do actually occur with this precise formula in the solid state, metal(III) hydroxides readily lose water on drying and usually occur in the solid state as $MO(OH)$. The compound with the formula $FeO(OH)$, for example, is 'rust'.

As more sodium hydroxide is added, further H_3O^+ is removed and blue copper(II) hydroxide precipitates. Thus, the overall equation is:

$$[Cu(H_2O)_6]^{2+} + 2OH^- \rightleftharpoons [Cu(OH)_2(H_2O)_4] + 2H_2O$$

The same sequence occurs whatever the metal ion, so that the reactions of sodium hydroxide with transition-metal ions can be summarised as:

- The addition of *sodium hydroxide* to a solution of a transition-metal salt *always precipitates* the transition-metal hydroxide.

The colours of some common hydroxides formed in these reactions are:

$$[Fe(H_2O)_6]^{2+} \qquad \rightarrow \qquad [Fe(H_2O)_4(OH)_2]$$

green solution green precipitate

$$[Fe(H_2O)_6]^{3+} \qquad \rightarrow \qquad [Fe(H_2O)_3(OH)_3]$$

pale violet solution brown precipitate

The addition of aqueous ammonia to a solution of a transition-metal salt results in exactly the same initial sequence of events as occur on the addition of sodium hydroxide. The ammonia upsets the hydrolysis equilibrium by removing H_3O^+ until the metal hydroxide is formed:

$$NH_3 + H_3O^+ \rightarrow NH_4^+ + H_2O$$

- The addition of *ammonia solution* to a solution of a transition-metal salt *always* results, *initially*, in the formation of a *precipitate* of the metal hydroxide.

A further reaction (i.e. *substitution*; see this book, section 3.2.5.2) can occur in the presence of an excess of ammonia, but initially a *precipitate* of the metal hydroxide is always formed.

Reactions of aqua ions with carbonate ions

Sodium carbonate is commonly used as a base, so it is important to be aware of its reactions with metal ions. The carbonate ion reacts with acids to form the hydrogencarbonate ion and then carbonic acid (H_2CO_3), which rapidly becomes carbon dioxide and water:

$$CO_3^{2-} + H_3O^+ \rightleftharpoons HCO_3^- + H_2O$$

$$HCO_3^- + H_3O^+ \rightleftharpoons CO_2 + 2H_2O$$

Metal(II) ions are not sufficiently acidic to displace carbonic acid from carbonates. Instead, these ions form insoluble metal carbonates, $MCO_3(s)$.

- The addition of *sodium carbonate solution* to a solution of a metal(II) salt results in a *precipitate* of the metal(II) carbonate.

Metal(III) ions, however, are more acidic than carbonic acid and therefore displace this acid from solutions containing carbonate ions (a strong acid will always displace a weaker one). So the reaction between a solution of metal(III) ions and carbonate ions results in *effervescence* of carbon dioxide and *precipitation* of the metal hydroxide.

The reaction follows the scheme:

$$[M(H_2O)_6]^{3+} + H_2O \rightleftharpoons [M(H_2O)_5(OH)]^{2+} + H_3O^+$$

followed by the reaction of H_3O^+ with CO_3^{2-}:

$$2H_3O^+ + CO_3^{2-} \rightleftharpoons CO_2 + 3H_2O$$

so that the equilibrium is pushed to the right and subsequent equilibria are set up:

$$[M(H_2O)_5(OH)]^{2+} + H_2O \rightleftharpoons [M(H_2O)_4(OH)_2]^+ + H_3O^+$$

$$[M(H_2O)_4(OH)_2]^+ + H_2O \rightleftharpoons [M(H_2O)_3(OH)_3] + H_3O^+$$

until the metal(III) hydroxide precipitates.

In general:

- The reaction between a metal(III) salt solution and sodium carbonate leads to *evolution of a gas* (carbon dioxide) and *precipitation* of the metal(III) hydroxide.

As a result, metal(III) carbonates cannot be prepared in aqueous solution and, in general, they do not exist. So, for example, $Al_2(CO_3)_3$ and $Fe_2(CO_3)_3$ *cannot* be prepared in aqueous solution and are, indeed, unknown.

The three equations in the hydrolysis of a metal(III) ion can be added together to give the overall equation:

$$[M(H_2O)_6]^{3+} + 3H_2O \rightleftharpoons [M(H_2O)_3(OH)_3] + 3H_3O^+$$

This equilibrium is driven to the right following removal of H_3O^+ by carbonate ions:

$$2H_3O^+ + CO_3^{2-} \rightarrow CO_2 + 3H_2O$$

Combining these two equations gives:

$$2[M(H_2O)_6]^{3+} + 3CO_3^{2-} \rightarrow 2[M(H_2O)_3(OH)_3] + 3CO_2 + 3H_2O$$

Amphoteric character

All of the hydrolysis equilibria given above can be reversed by using strong acids. Thus, metal hydroxides dissolve in strong acids to give metal-aqua ions (as long as a non-complexing strong acid such as HNO_3 is used; nitrate ions are only *weak* ligands).

Whilst metal hydroxides do not undergo hydrolysis in water, they can be attacked by the strong nucleophile OH^- to give anionic complexes which are water-soluble. This ability of metal hydroxides to dissolve in both strong acids and strong alkalis is known as *amphoteric character* (see this book, section 3.2.4).

When sodium hydroxide is added to a solution of an aluminium salt, a white precipitate of aluminium hydroxide is seen initially. Then, as an excess of the alkali is added, the precipitate dissolves to form a colourless solution containing the aluminate ion:

$$[Al(H_2O)_6]^{3+} + 3H_2O \rightleftharpoons [Al(H_2O)_3(OH)_3] + 3H_3O^+$$

$$[Al(H_2O)_3(OH)_3] + OH^- \rightleftharpoons [Al(OH)_4]^- + 3H_2O$$

The general scheme for metal(III) ions is:

$$[M(H_2O)_6]^{3+} \underset{H_3O^+}{\overset{OH^-}{\rightleftharpoons}} [M(H_2O)_3(OH)_3] \underset{H_3O^+}{\overset{OH^-}{\rightleftharpoons}} [M(OH)_4]^-$$

acidic solution   neutral solution   alkaline solution

Most transition-metal hydroxides are amphoteric.

Notes

It is not only carbonate ions which behave in this way. Anions derived from weak acids are protonated by metal(III)-aqua ions so that the free acid and the metal(III) hydroxide result.

Some examples of anions of weak acids are S^{2-}, NO_2^-, SO_3^{2-}, $S_2O_3^{2-}$ and CH_3COO^-. Metal(III) salts of these anions cannot usually be prepared in aqueous solution.

Notes

Whereas *all* bases yield hydroxide precipitates, only *strong bases in excess* can cause some $M(H_2O)_3(OH)_3$ precipitates to re-dissolve, forming negatively-charged complexes. An excess of a *weak base* does not re-dissolve these hydroxide precipitates.

A summary of the reactions of metal(II) and metal(III) ions with various bases (hydroxide ions, ammonia and carbonate ions) is given in Tables 25 and 26.

Table 25
Summary of the reactions of metal(II) ions with bases

Base added	Aqueous M(II) ions	
	$[Fe(H_2O)_6]^{2+}$ green solution	$[Cu(H_2O)_6]^{2+}$ blue solution
OH^- (little)	$[Fe(H_2O)_4(OH)_2]$ green ppt	$[Cu(H_2O)_4(OH)_2]$ blue ppt
OH^- (excess dilute)	(turns brown in air)	
NH_3 (little)		
NH_3 (excess)	as above	$[Cu(NH_3)_4(H_2O)_2]^{2+}$ deep-blue solution
CO_3^{2-}	$FeCO_3$ green ppt	$CuCO_3$ green-blue ppt

Notes

Carrying out of simple test-tube reactions to identify transition metal ions in aqueous solution is a required practical activity.

Table 26
Summary of the reactions of metal(III) ions with bases

Base added	Aqueous M(III) ions	
	$[Fe(H_2O)_6]^{3+}$ violet solution (appears yellow due to hydrolysis)	$[Al(H_2O)_6]^{3+}$ colourless solution
OH^- (little)	$[Fe(H_2O)_3(OH)_3]$ brown ppt	$[Al(H_2O)_3(OH)_3]$ white ppt
OH^- (excess)	as above brown ppt	$[Al(OH)_4]^-$ colourless solution
NH_3 (little)	as above brown ppt	$[Al(H_2O)_3(OH)_3]$ white ppt
NH_3 (excess)	as above brown ppt	as above white ppt
CO_3^{2-}	as above brown ppt and effervescence of CO_2	as above white ppt and effervescence of CO_2

Practical and mathematical skills

In the A-level Paper 1, approximately 15% of marks will be allocated to the assessment of skills related to practical chemistry. A minimum of 20% of the marks will be allocated to assessing level 2 mathematical skills. These practical and mathematical skills are likely to overlap.

The required practical activities assessed in this paper are:

- Make up a volumetric solution and carry out a simple acid–base titration.
- Measurement of an enthalpy change.
- Carry out simple test-tube reactions to identify cations (Group 2 and NH_4^+) and anions (halide, hydroxide, carbonate and sulfate ions).
- Measuring the EMF of an electrochemical cell.
- Investigate how pH changes when a weak acid reacts with a strong base and when a strong acid reacts with a weak base.
- Carry out simple test-tube reactions to identify transition metal ions in aqueous solution.

The practical skills assessed in the paper are:

1. Independent thinking

Examination questions may require problem solving and the application of scientific knowledge and understanding in practical contexts. For example, a question may ask how an experiment could be carried out to investigate the effect on the EMF of an electrochemical cell of changes in concentration or of temperature. Another example is a question that requires the use of data from a curve, drawn using pH values measured during the titration of a weak dibasic acid with a strong base, to select an appropriate indicators for each end-point.

2. Use and application of scientific methods and practices

This skill may be assessed by asking for critical comments on a given experimental method. Questions may require the presentation of data in appropriate ways, such as in tables or graphs. It will also be necessary to express numerical results, for example from titration and enthalpy determinations, to an appropriate precision with reference to uncertainties and errors in volumetric apparatus and in thermometer readings. Questions may ask for conclusions to be made from observations, for example in the test-tube reactions of transition metal ions in aqueous solution with sodium hydroxide, ammonia and sodium carbonate.

3. Numeracy and the application of mathematical concepts in a practical context

There is some overlap between this skill and the use and the application of scientific methods and practices. Questions may require the plotting of pH curves for the titration of a weak acid with a strong base and the use of this curve to determine the pK_a value of the acid. It will also be necessary write the conventional representation of an electrochemical cell and to calculate the EMF.

4. Instruments and equipment

It will be necessary to know and understand how to standardise a pH meter, how to use it and volumetric glassware to determine pH values during acid–base titrations. It may also be necessary to explain how to set up electrochemical cells and measure the EMF of a range of cells. Questions will assess the ability to understand in detail how to make appropriate observations of test-tube reactions of Group 2 cations and transition metal ions in aqueous solution and how to draw valid conclusions from these observations.

The mathematical skills assessed in this paper are:

1. Arithmetic and numerical computation

* **Recognise and make use of appropriate units in calculations.**

  All numerical answers should be given with the appropriate units. Questions may require conversions between units, for example, cm^3 to dm^3 and J to kJ.

* **Recognise and use expressions in decimal and standard form.**

  When required, it will be necessary to express answers to an appropriate number of decimal places and to carry out calculations and express answers in ordinary or standard form. For example, calculations involving concentrations may involve numbers in standard form and conversions between standard and ordinary form. The appropriate number of decimal places must be used in pH calculations. Other calculations that require the use of both standard and ordinary form include pH calculations and those using the Avogadro constant.

* **Use ratios fractions and percentages.**

  Example of this skill include the calculation of percentage yields, atom economies and the construction and/or balancing of equations using ratios.

* **Estimate results.**

  Estimations of this type could include the evaluation of how the addition of a small quantity of acid or alkali would change the pH of a buffer solution.

* **Use calculators.**

  The ability to use calculators to find power, exponential and logarithmic functions and to use these functions in pH, pK_a and Avogadro constant calculations and to make appropriate approximations in buffer calculations.

2. Handling data

* **Use an appropriate number of significant figures.**

  Understand that a calculated result can only be reported to the limits of the least accurate measurement, for example in the selection of appropriate readings to determine a mean titre in a volumetric titration or the measurement of temperature in an enthalpy of combustion experiment.

- **Find arithmetic means.**

  Examples may include the determination of the mean bond enthalpy from data for a given bond enthalpy in a range of compounds.

- **Identify uncertainties in measurements and when data are combined.**

  It will be necessary to demonstrate an ability to determine uncertainty when two burette readings are used to calculate a titre value in a volumetric analysis or when a temperature difference is calculated from two thermometer readings in an experiment to determine an enthalpy change.

3. **Algebra**

- **Change the subject of an equation.**

  For example when the concentration of hydrogen ions in an aqueous solution of a weak acid is calculated from the concentration of the acid and the acid dissociation constant. Another example is the rearrangement of the equation: $\Delta G = \Delta H - T\Delta S$ in order to find the temperature at which a reaction becomes feasible.

4. **Graphs**

- **Plot two variables from experimental data.**

  Examples of this skill include the plotting of pH curves using data obtained from a pH meter during a range of acid–base titrations. The use of these pH curves to determine the end-point of the reaction, the selection of suitable indicators for the determination of the end-point and for the determination of the pK_a values for weak acids.

5. **Geometry and trigonometry**

- **Visualise and represent 2D and 3D forms.**

  Questions may assess an ability to predict, identify and sketch the shapes of and predict bond angles in molecules and ions with and without lone pairs, for example NH_3, $AlCl_4^-$, etc. Questions may also ask for 2D and 3D representations of octahedral, tetrahedral, square-planar and linear complexes formed by transition metal ions. These include E and Z isomers of octahedral complexes formed by transition metal ions with monodentate ligands and optical isomers formed with bidentate ligands. Questions may also ask for the bond angles in these complexes to be predicted.

Practice exam-style questions

1   In the Contact process for the production of sulfuric acid, sulfur dioxide and oxygen react to form sulfur trioxide in the presence of a heterogeneous catalyst. Temperatures in excess of 500 K and pressures in the region of 100 – 200 kPa are commonly used. The equation for the reaction is:

$$2SO_2(g) + O_2(g) \rightleftharpoons 2SO_3(g) \qquad \Delta H^{\ominus} = -190 \text{ kJ mol}^{-1}$$

(a)   Identify a catalyst that is used in industry for this reaction and explain what is meant by *heterogeneous* in this context.

Catalyst _____

Meaning of heterogeneous _____

_____ 2 marks

(b)   The gases for the process are purified before they are passed over the catalyst. Suggest a reason why this is done even though it adds to the cost of the process.

_____

_____ 1 mark

(c)   The sulfur dioxide for the process is often made by burning sulfur in air. State what is observed when sulfur burns in air and give an equation for the reaction.

Observations _____

_____

Equation _____ 3 marks

(d)   In an experiment to determine the value of K_p for the equilibrium above at a certain temperature, T_1, 1.30 mol of sulfur dioxide were mixed with 0.63 mol of oxygen. At equilibrium, 0.21 mol of sulfur dioxide remained. The pressure of the equilibrium mixture was 110 kPa.

(i)   Deduce the amounts, in moles, of the oxygen and sulfur trioxide in the equilibrium mixture.

Amount of oxygen _____

Amount of sulfur trioxide _____ 2 marks

(ii)   Calculate the mole fractions of the sulfur dioxide, oxygen and sulfur trioxide in the equilibrium mixture.

Mole fraction of sulfur dioxide _____

Mole fraction of oxygen _____

Mole fraction of sulfur trioxide _____ 1 mark

(iii)   Calculate a value for the equilibrium constant, K_p, for the Contact Process at temperature T_1, giving your answer to an appropriate number of significant figures. Give the units for K_p.

_____

_____

_____

_____

_____
_____
_____
_____
_____
_____ 5 marks

(e) In a different experiment at temperature, T_2, the value of the equilibrium constant, K_p, was found to be 3.8×10^{-3}. State whether T_2 is at a higher or lower temperature than T_1 and explain your reasoning.

_____
_____
_____ 3 marks

Total marks: 17

2 (a) Explain what is meant by *atomisation enthalpy*.

_____
_____ 2 marks

(b) The diagram below is an incomplete Born–Haber cycle for the formation of calcium oxide. The diagram is not to scale.

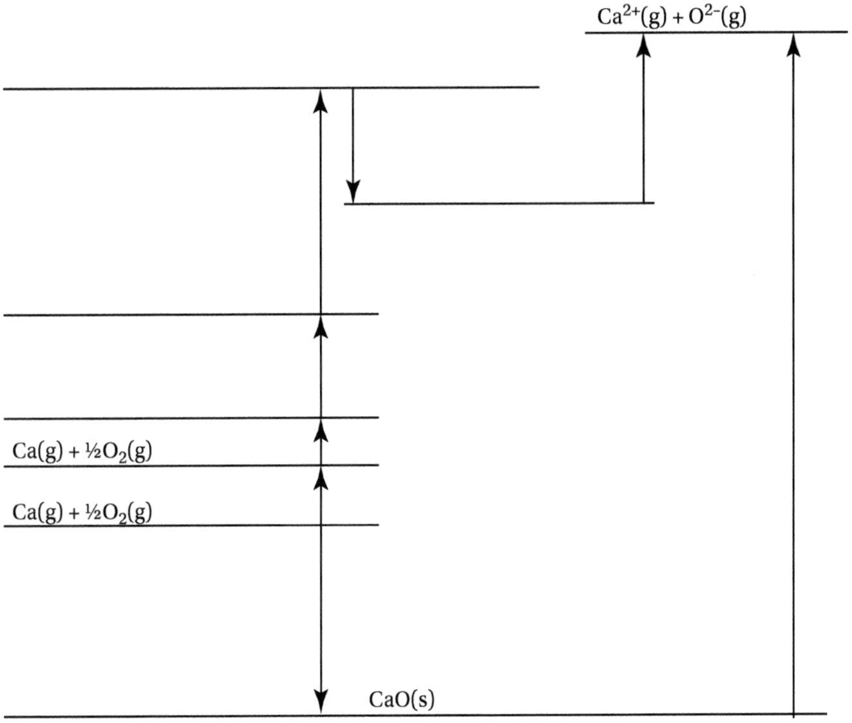

$$Ca^{2+}(g) + O^{2-}(g)$$

$$Ca(g) + \tfrac{1}{2}O_2(g)$$

$$Ca(g) + \tfrac{1}{2}O_2(g)$$

CaO(s)

(i) Complete the diagram by writing the appropriate chemical symbols, with state symbols, on each of the four blank lines.

4 marks

(ii) Use the enthalpy changes in the table below to calculate a value for the second electron affinity for oxygen.

3 marks

Name of enthalpy change	$\Delta H^{\ominus}$/kJ mol^{-1}
Bond dissociation enthalpy of oxygen	+498
First electron affinity of oxygen	−142
Enthalpy of atomisation of calcium	+178
First ionisation enthalpy of calcium	+732
Second ionisation enthalpy of calcium	+1150
Enthalpy of formation of calcium oxide	−635
Enthalpy of lattice formation of calcium oxide	−3646

(iii) Explain why the first electron affinity for oxygen is exothermic.

1 mark

(iv) Explain why the second electron affinity for oxygen is endothermic.

2 marks

Total marks: 12

3 (a) Explain what is meant by the term *ionic model*.

2 marks

(b) Values for the theoretical lattice formation enthalpies (calculated using the ionic model) and the experimental value (using Born–Haber cycles) are given below for some chlorides.

Compound	Theoretical lattice enthalpy/kJ mol^{-1}	Experimental lattice enthalpy/kJ mol^{-1}
NaCl	−766	−771
KCl	−705	−715
AgCl	−770	−905

(i) Explain the difference between the theoretical value of the lattice enthalpy for NaCl and that for KCl.

_____

_____ 2 marks

(ii) What can be deduced about the bonding in sodium chloride from the values? Explain your answer.

_____

_____ 2 marks

(iii) What can be deduced about the bonding in silver chloride from the values? Explain your answer.

_____

_____ 2 marks

Total marks: 8

4 Use the following data to help you answer this question.

$$Mg^{2+}(g) \rightarrow Mg^{2+}(aq) \qquad \Delta H^{\ominus} = -1926 \text{ kJ mol}^{-1}$$
$$MgCl_2(s) \rightarrow Mg^{2+}(aq) + 2Cl^-(aq) \qquad \Delta H^{\ominus} = -155 \text{ kJ mol}^{-1}$$
$$Mg^{2+}(g) + 2Cl^-(g) \rightarrow MgCl_2(s) \qquad \Delta H^{\ominus} = -2493 \text{ kJ mol}^{-1}$$

(a) (i) Explain what is meant by the term *enthalpy of hydration*, $\Delta_{hyd}H^{\ominus}$, of an ion.

_____

_____ 2 marks

(ii) Calculate the standard enthalpy of hydration of the chloride ion.

_____

_____

_____

_____ 3 marks

(b) The value of the enthalpy of hydration of the fluoride ion is -524 kJ mol^{-1} and that of the bromide ion is -348 kJ mol^{-1}. Explain the difference between the enthalpies of hydration of the fluoride ion and the bromide ion.

_____

_____

_____ 3 marks

Total marks: 8

5 Two students did an experiment to find the enthalpy of reaction between sodium hydrogencarbonate and dilute hydrochloric acid. The equation for the reaction is given below.

$$NaHCO_3 + HCl \rightarrow NaCl + H_2O + CO_2$$

They measured the temperature of 25.0 cm^3 of an aqueous solution of hydrochloric acid of concentration 1.10 mol dm^{-3} for three minutes. At the fourth minute they added 2.14 g of sodium hydrogencarbonate powder. They measured the temperature of the reaction mixture until ten minutes had elapsed. The results of their experiment are given in the table below.

Time/mins	0	1	2	3	4	5	6	7	8	9	10
Temperature/°C	18.2	18.1	18.2	18.2	*	13.5	11.9	12.1	12.3	12.4	12.7

(a) Label the axes and plot a graph of their results. Draw two lines of best fit and use your graph to determine the temperature change at the fourth minute.

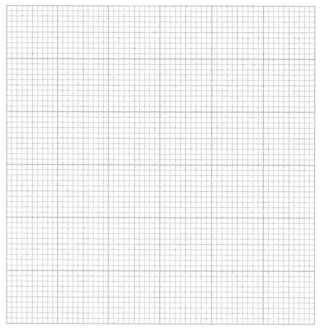

Temperature change at the fourth minute _____ 5 marks

(b) Calculate the molar enthalpy change for the reaction. The specific heat capacity of the reaction mixture is 4.2 J g^{-1} K^{-1}. Assume in your calculation that the mass of the reaction mixture is 25.0 g.

Give your answer to an appropriate number of significant figures.

_____

_____

_____

_____

_____

_____

_____

_____

_____

_____

_____ 5 marks

(c) **(i)** One of the students suggested that the percentage uncertainty in measuring the temperature change was the greatest source of error. Use the following data to decide whether the student was correct and suggest how this percentage uncertainty could be reduced and justify your answer.

The temperature was measured with a thermometer with a total uncertainty of $\pm0.2\,°C$

The mass of sodium hydrogencarbonate was measured with a balance with a total uncertainty of $\pm0.01\,g$

The volume of the hydrochloric acid was measured with a pipette with a total uncertainty of $\pm0.1\,cm^3$

_____

_____

_____

_____

_____

_____ 5 marks

(ii) Explain briefly why the *concentration* of the hydrochloric acid does not affect the overall error in this determination.

_____

_____ 2 marks

(d) The standard enthalpy change of the reaction between sodium hydrogencarbonate and dilute hydrochloric acid is given in the literature as $+28.5\,kJ\,mol^{-1}$.

(i) Calculate the difference between the value you found in (c)(i) and this value and express this as a percentage of the literature value.

_____

_____ 1 mark

(ii) Explain why this reaction is spontaneous at these temperatures despite the sign of the enthalpy change.

_____

_____

_____

_____

_____

_____ 5 marks

Total marks: 23

6 The feasibility of a chemical reaction depends on the standard free-energy change, $\Delta G^\ominus$.

   (a) Write an equation that relates $\Delta G^\ominus$ to the standard enthalpy change, $\Delta H^\ominus$, and the standard entropy change, $\Delta S^\ominus$.

_____ 1 mark

   (b) In terms of $\Delta G^\ominus$, state the necessary condition for a feasible reaction.

_____ 1 mark

   (c) At a pressure of 100 kPa, water will not freeze while the surrounding temperature remains above 0 °C.

     (i) State the sign of the enthalpy change and of the entropy change during freezing.

        Sign of $\Delta H^\ominus$ _____

        Sign of $\Delta S^\ominus$ _____ 2 marks

     (ii) Explain, in terms of $\Delta G^\ominus$, why water does not freeze at temperatures above 0 °C.

_____

_____ 2 marks

   (d) Potassium chlorate(V) decomposes on heating according to the equation:

$$4KClO_3(s) \rightarrow 3KClO_4(s) + KCl(s) \qquad \Delta H^\ominus = +18 \text{ kJ mol}^{-1}$$

The table below shows the standard enthalpies of formation, $\Delta_f H^\ominus$, and the standard molar entropy, $S^\ominus$, for the species involved in the reaction.

	$KClO_3(s)$	$KClO_4(s)$	$KCl(s)$
$\Delta_f H^\ominus/\text{kJ mol}^{-1}$	−391	to be determined	−436
$S^\ominus/\text{kJ mol}^{-1}$	112	134	83

     (i) Calculate the enthalpy of formation of $KClO_4(s)$. Give the units.

_____

_____

_____ 3 marks

     (ii) Calculate the standard entropy change for the decomposition of $KClO_3$ according to the equation above. Give the units.

_____

_____

_____ 3 marks

(iii) Calculate the standard free-energy change at 298 K for this decomposition. Give the units.

_____

_____

_____ 3 marks

(iv) At what temperature does the thermal decomposition of potassium chlorate(V) become feasible?

_____

_____

_____ 3 marks

Total marks: 18

7   The following equation shows the formation of water.

$$\frac{1}{2}O_2(g) + H_2(g) \rightarrow H_2O(g)$$

The graph shows how the standard free-energy change for this reaction varies with temperatures above 4000 K.

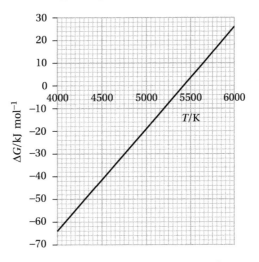

(a) Write an equation to show the relationship between $\Delta G^\ominus$, $\Delta H^\ominus$ and $\Delta S^\ominus$.

_____ 1 mark

(b) Use the graph to calculate a value for the slope (gradient) of the line. Give the units of this slope and the symbol for the thermodynamic quantity that this slope represents.

Value of the slope _____

Units _____

Symbol _____ 3 marks

(c) Explain the significance, for this reaction, of temperatures below the temperature value where the line crosses the temperature axis.

_____

_____ 2 marks

(d) Explain why the complete combustion of methanol is spontaneous at and above 373 K.

The boiling point of methanol is 338 K.

_____

_____

_____

_____

_____

_____

_____

_____

_____

_____

_____

_____

_____

_____

_____

_____

_____ 6 marks

Total marks: 12

8 **(a)** In terms of electron transfer, what is the role of a reducing agent?

_____ 1 mark

(b) Standard electrode potentials, $E^{\ominus}$, are measured relative to a standard reference electrode.

(i) Name the standard reference electrode and state its potential. State why this is the potential assigned to this electrode.

Standard reference electrode _____

Electrode potential _____

Explanation _____ 3 marks

(ii) State **three** conditions that must apply when values of $E^{\ominus}$ are being measured using this standard reference electrode.

Condition 1 _____

Condition 2 _____

Condition 3 _____ 3 marks

(iii) A *salt bridge* is used when making such measurements. Explain the function of a *salt bridge*, and state what chemical substance it might contain.

Function _____

Contents _____ 2 marks

(c) Explain the term *electrochemical series*.

_____

_____ 2 marks

(d) The standard electrode potentials, $E^{\ominus}$, for two electrodes are shown below. Use this information to answer the questions that follow.

Iron electrode immersed in a solution of aqueous iron(II) ions $E^{\ominus}/V = -0.44$

Zinc electrode immersed in a solution of aqueous zinc(II) ions $E^{\ominus}/V = -0.76$

(i) Write the half-equation that applies to the standard electrode potential for zinc

_____ 1 mark

(ii) The two electrodes and their solutions are used to form an electrochemical cell. Calculate the cell EMF.

Cell EMF _____ 1 mark

(iii) State which species is reduced when the two electrodes are connected together. Explain your answer.

Species reduced _____

Explanation _____ 2 marks

Total marks: 15

9 Electrochemical cells can be used as commercial sources of electrical energy.

(a) The lithium cell is classed as *disposable* and has a relatively long life. The materials used are lithium metal and manganese(IV) oxide, both of which are fairly inexpensive. In this cell, the overall reaction is:

$$Li(s) + MnO_2(s) \rightarrow LiMnO_2(s)$$

(i) Explain what, in this context, is meant by the term *disposable*.

_____ 1 mark

(ii) Give **one** benefit and **one** risk to society of the use of such cells.

Benefit _____

Risk _____

_____ 2 marks

(iii) Given that one of the reactions in this cell involves the oxidation of lithium, write the half-equations for the oxidation and reduction reactions in the lithium cell. State the oxidation state of manganese in $LiMnO_2$

Oxidation reaction _____

Reduction reaction _____

Oxidation state of Mn in $LiMnO_2$ _____ 3 marks

(b) The cell reaction in the *dry cell* (Leclanché cell) is:

$$Zn(s) + 2MnO_2(s) + 4NH_4^+(aq) + 2OH^-(aq) \rightarrow [Zn(NH_3)_4]^{2+}(aq) + 2MnO(OH)(s) + 2H_2O(l)$$

This cell has EMF = 1.51 V and a standard enthalpy change for the cell reaction, $\Delta H^{\ominus} = -263$ kJ mol^{-1}.

(i) Identify the oxidising agent in this reaction and give the oxidation states of the element that is reduced before and after the reaction.

Oxidising agent _____

Oxidation state of element before reaction _____

Oxidation state of element after reaction _____ 3 marks

(ii) This cell is not rechargeable. Suggest why this cell cannot be recharged.

_____ 1 mark

(c) The lead–acid battery is commonly used as a source of electrical energy in motor vehicles. The overall cell reaction and one of the two half-equations in this cell are shown below.

The EMF of the cell is +2.04 V.

Overall cell equation: $PbO_2(s) + Pb(s) + 2H^+(aq) + 2HSO_4^-(aq) \rightleftharpoons 2PbSO_4(s) + 2H_2O(l)$

One half-equation: $PbO_2(s) + 3H^+(aq) + HSO_4^-(aq) + 2e^- \rightarrow PbSO_4(s) + 2H_2O(l)$ $E^{\ominus} = +1.68$ V.

(i) Explain the significance of the equilibrium arrows in the overall cell equation

_____ 1 mark

(ii) Deduce the other half-equation in this cell and calculate its standard electrode potential.

Half-equation _____

Value of $E^{\ominus}$ _____ 2 marks

(d) Commercial hydrogen–oxygen fuel cells are operated with an electrolyte of concentrated aqueous potassium hydroxide at a temperature of 200 °C under a pressure of 3000 kPa.

(i) Write a half–equation for the reaction of hydrogen with hydroxide ions to form water that occurs at one of the porous electrodes in this electrodes in this fuel cell. The standard electrode potential for this process is –0.83 V.

Half-equation _____ 1 mark

(ii) Write the half-equation for the reaction of one mole of gaseous oxygen with water to form aqueous hydroxide ions at the other porous electrode in this fuel cell. The standard electrode potential for this process is +0.40 V.

Half-equation _____ 1 mark

(iii) Hence deduce the overall equation for the reaction in this fuel cell and calculate the cell EMF.

Overall equation _____

Cell EMF _____ 2 marks

(iv) The fuel cell may also be run under acid conditions for which the half-equations are:

$$2H^+(aq) + 2e^- \rightarrow H_2$$
$$O_2 + 4H^+ + 4e^- \rightarrow 2H_2O$$

Deduce the overall cell equation and hence the $E^{\ominus}$ for O_2/H_2O in acid conditions.

Overall cell equation _____

$E^{\ominus}(O_2/H_2O)$ _____ 2 marks

(v) On the axes below, sketch a graph of how the output voltage of this fuel cell changes with time when the cell is in normal use. Explain why the graph has this form.

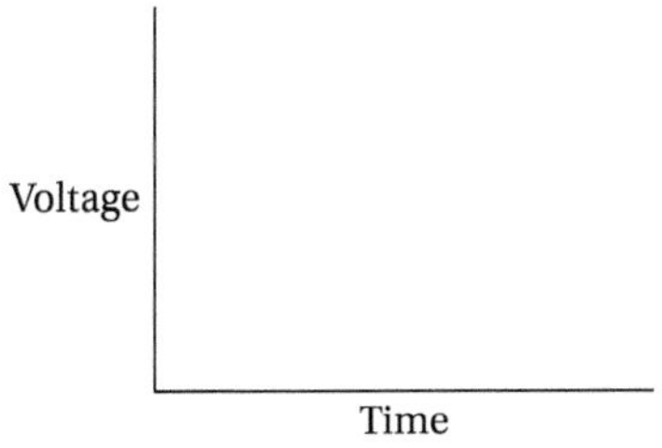

Voltage |

Time

Explanation _____

_____

_____ 2 marks

Total marks: 21

10 (a) Nitric acid is a strong Brønsted–Lowry acid.

(i) Explain what is meant by the terms *Brønsted–Lowry acid* and *strong acid*.

Brønsted Lowry acid _____

Strong acid _____ 2 marks

(ii) Calculate the pH of an aqueous solution of nitric acid with concentration 1.19 mol dm^{-3}.

pH _____ 1 mark

(b) Some values of the ionic product of water, K_w, are shown in the table below.

Temperature/K	Value of K_w
273	1.14×10^{-15}
323	5.48×10^{-14}
373	5.13×10^{-13}

(i) Define the term *ionic product of water*, K_w, and give its value at 298 K, with units.

Definition of K_w _____

Value at 298 K_w _____

Units of K_w _____ 3 marks

(ii) Write an equation for the dissociation of water.

_____ 1 mark

(iii) Calculate the pH of pure water at 373 K.

_____

_____

_____ 3 marks

(iv) Explain why pure water at 373 K is neutral even though the pH is not 7.00.

_____ 1 mark

(v) Use the values of K_w to deduce the enthalpy change for the dissociation of water. Explain your reasoning.

Sign of enthalpy change

Explanation

_____ 2 marks

(c) **(i)** Calculate the pH of a 0.131 mol dm^{-3} aqueous solution of sodium hydroxide at 323 K.

_____ 2 marks

(ii) Calculate the pH of the solution formed when 12.5 cm^3 of a 0.140 mol dm^{-3} aqueous solution of sulfuric acid is added to 25.0 cm^3 of the sodium hydroxide in (c)(i). Assume the sulfuric acid is fully ionised.

_____ 5 marks

Total marks: 20

11 A student performed an experiment to measure the pH of the solution during the titration of 25.0 cm^3 of a dilute aqueous solution of butanoic acid of unknown concentration with 0.100 mol dm^{-3} sodium hydroxide. In the experiment, portions of sodium hydroxide were added to the butanoic acid and the pH of the mixture recorded after each addition.

The results are shown below:

Volume of NaOH added/cm³	0.0	2.0	4.0	6.0	8.0	10.0	12.0	14.0	16.0
pH of mixture		3.8	4.2	4.2	4.6	4.7	4.9	5.1	5.1

Volume of NaOH added/cm³	18.0	20.0	22.0	24.0	26.0	28.0	30.0	35.0	40.0
pH of mixture	5.5	5.8	9.2	11.6	11.9	12.1	12.2	12.3	12.4

(a) **(i)** Label the axes. Plot a graph of the points and draw a best-fit curve.

4 marks

(ii) Use your graph to determine the volume of sodium hydroxide needed to react exactly with the butanoic acid and hence find the concentration of the original butanoic acid solution.

Volume of sodium hydroxide _____

Concentration of butanoic acid _____

_____

_____ 3 marks

(iii) The pK_a of butanoic acid is 4.83. Use this datum and your answer in part (a)(ii) to calculate the pH of the acid solution before the addition of the sodium hydroxide. Add your answer to the graph.

_____

_____

_____ 3 marks

(b) Use your graph to find the pH of the solution formed when half the acid has reacted. Use this value to calculate the value of K_a for butanoic acid.

_____

_____

_____

_____ 3 marks

(c) The experiment was repeated but using hydrochloric acid of the same concentration as the butanoic acid.

    (i) State two features (such as volume, pH value or shape) that are exactly the same in the two titration curves.

        Identical feature 1 _____

        Identical feature 2 _____ 2 marks

    (ii) State two features (such as volume, pH value or shape) that are distinctly different in the titration curves.

        Different feature 1 _____

        Different feature 2 _____ 2 marks

(d) A table of indicators with their pH ranges is given below.

Name of indicator	pH range
Methyl violet	0.0 – 2.0
Bromophenol blue	3.0 – 4.6
Phenol red	6.4 – 8.0
Cresolphthalein	8.2 – 9.8
Alizarin Yellow R	10.2 – 12.0

    (i) Identify an indicator which could be used for the titration of butanoic acid and hydrochloric acid.

    _____ 1 mark

    (ii) Identify an indicator which could be used for the titration of hydrochloric acid but not butanoic acid.

    _____ 1 mark

(e) The pH was measured in this experiment with a pH meter. This must be calibrated before the experiment. Describe how this could be done and explain why calibration is necessary.

_____

_____

_____

_____ 3 marks

_____

Total marks: 22

12 (a) Describe what is meant by the term *buffer solution*.

_____

_____

_____ 2 marks

(b) (i) State what is meant by the term *acidic buffer*.

_____ 1 mark

(ii) Give two reagents you could use to make such a buffer. Write an equation for the equilibrium established in a buffer solution containing these reagents.

Reagents _____

Equation _____ 2 marks

(iii) Use this equation to illustrate how the acidic buffer works.

_____

_____

_____

_____ 4 marks

(c) Calculate the pH of the buffer solution that results from the addition of 12.0 cm^3 of 0.20 mol dm^{-3} NaOH to 20.0 cm^3 of 0.24 mol dm^{-3} of a weak acid HX (pKa = 5.00).

_____

_____

_____

_____

_____

_____

_____

_____

_____

_____

_____

_____ 6 marks

Total marks: 15

13 (a) Explain, in terms of their bonding and structures, why the melting point of magnesium oxide is much higher than that of phosphorus(V) oxide.

_____

_____

_____

_____

_____ 6 marks

(b) State the observations you would make and write an equation for any reaction occurring, when samples of magnesium oxide and phosphorus(V) oxide were to be added separately to water. Give the final pH of each of the mixtures.

Observation with magnesium oxide _____

Equation with magnesium oxide _____

pH of mixture _____

Observation with phosphorus(V) oxide _____

Equation with phosphorus(V) oxide _____

pH of mixture _____ 6 marks

(c) Aluminium and silicon both react with oxygen to form oxides.

(i) Write an equation for the reaction of aluminium with oxygen. Suggest why this reaction does not occur readily with a normal sample of aluminium.

Equation _____

Explanation _____

_____ 2 marks

(ii) Explain why aluminium oxide and silicon dioxide are both insoluble in water.

_____

_____

_____

_____ 4 marks

(iii) Give ionic equations to show why aluminium oxide and silicon dioxide both dissolve in aqueous sodium hydroxide.

Equation for aluminium oxide _____

Equation for silicon dioxide _____ 2 marks

Total marks: 20

14 (a) Deduce the oxidation state and co-ordination number of chromium in the complex compound $[Cr(H_2O)_4F_2]F$.

Oxidation state of chromium _____

Co-ordination number of chromium _____ 2 marks

(b) When thiocyanate ions, NCS^-, are added to aqueous iron(III) ions a reaction occurs as shown in the equation below.

$$[Fe(H_2O)_6]^{3+} + NCS^- \rightarrow [Fe(SCN)(H_2O)_5]^{2+} + H_2O$$

(i) Give the electronic configurations and co-ordination numbers of the iron ions in each of the complexes

_____

_____ 2 marks

(ii) The initial complex, $[Fe(H_2O)_6]^{3+}$, is pale violet in colour. Explain the origin of the colour in this complex.

_____

_____

_____

_____ 3 marks

(iii) The maximum absorption of the $[Fe(SCN)(H_2O)_5]^{2+}$ complex occurs at an energy of 4.139×10^{-19} J. Calculate the frequency of visible light corresponding to this energy, giving your answer to an appropriate precision. Include units in your answer.

The Planck constant has a value of 6.62607×10^{-34} J s.

_____

_____

_____ 2 marks

(c) Aqueous iron(III) ions undergo a reaction with ethanedioate ions. The ethanedioate ion acts as a bidentate ligand in this reaction.

$$[Fe(H_2O)_6]^{3+} + 3C_2O_4^{2-} \rightarrow [Fe(C_2O_4)_3]^{3-} + 6H_2O$$

(i) Explain what is meant by the term *bidentate ligand.*

_____

_____ 2 marks

(ii) State the type of reaction that occurs.

_____ 1 mark

(iii) The product exists as two stereoisomers. Draw the structures of the two stereoisomers showing how they are related. Show the structure of the ligand in your answer.

Isomer 1   Isomer 2   3 marks

Total marks: 15

15 Use your understanding of transition-metal chemistry and of aqueous metal ions to answer this question.

(a) Titanium(IV) chloride is a colourless liquid which reacts with moisture in air to form titanium(IV) oxide and an acidic gas only.

Write an equation for the reaction of titanium(IV) chloride with water.

_____ 1 mark

(b) (i) Write the outer electronic configurations of titanium(IV) and of titanium(III).

Ti(IV) [Ar] _____

Ti(III) [Ar] _____ 2 marks

(ii) Titanium(IV) chloride is colourless, but titanium(III) chloride is a purple solid. Suggest why titanium(III) chloride is coloured but titanium(IV) chloride is colourless.

_____

_____ 2 marks

(iii) Suggest an instrumental method for determining the concentration of a solution of titanium(III) chloride and outline how you would carry out an experiment to determine the concentration of a solution of unknown concentration.

_____

_____

_____

_____

_____

_____

_____

_____ 6 marks

(c) Titanium(III) ions are oxidised to titanium(IV) ions by potassium manganate(VII) in acidic solution.

Writing titanium(III) ions as Ti^{3+} and titanium(IV) ions as Ti^{4+}, construct an ionic equation for this reaction.

_____

_____ 2 marks

(d) Predict what you would see if an aqueous solution of titanium(III) chloride was treated with aqueous sodium carbonate. Write an equation for the overall reaction occurring.

Observations _____

_____

Equation _____ 4 marks

Total marks: 17

16 (a) (i) When anhydrous copper(II) sulfate is added to water, a blue solution is formed. Write an equation for this reaction, showing the formula of the species responsible for the blue colour.

Equation _____ 1 mark

(ii) The addition of aqueous ammonia to the blue solution gives a pale blue precipitate, which dissolves to form a deep blue solution when an excess of aqueous ammonia is added. Give equations for the reaction to form the blue precipitate and a separate equation to show the conversion of the blue precipitate into the deep blue solution.

Equation for formation of pale blue precipitate

_____ 2 marks

Equation for conversion of pale blue precipitate into deep blue solution

_____ 2 marks

(iii) If ethane-1,2-diamine, $H_2NCH_2CH_2NH_2$, is added to this deep blue solution, a further reaction occurs. The ammonia ligands are replaced by ethane-1,2-diamine.

Write an equation for the reaction showing the structure of the product.

Explain why the products are more stable than the reactants despite the enthalpy change for the reaction being close to zero.

Equation _____

Explanation _____

_____

_____ 5 marks

127

(b) The ammonia complex, *cisplatin*, $Pt(NH_3)_2Cl_2$, has found a use in medicine.

    (i) State its use and give a risk associated with that use.

      Use of cisplatin _____

      Risk of its use _____

      _____

      _____ 2 marks

    (ii) *Cisplatin* is the *Z*-isomer. Draw the structure of *transplatin*, the *E*-isomer, using the symbol → to represent any co-ordinate bonds.

      2 marks

(c) A solution of a silver compound in aqueous ammonia finds a use as a reagent in a laboratory test in organic chemistry.

    (i) Give the formula of the silver-containing species present in this solution and state its bond angle.

      Formula _____

      Bond angle _____ 2 marks

    (ii) Name the organic functional group which reacts in the test.

      _____ 1 mark

    (iii) Write an equation to show the reaction occurring in the test. Use the symbol *R* as the non-functional part of the organic molecule in the equation.

      _____

      _____ 2 marks

Total marks: 19

17 In a student's experiment to find the formula of an iron(II) salt, $FeSO_4 \cdot xH_2O$, 6.45 g of the iron salt were dissolved in dilute sulfuric acid. The solution was transferred to a volumetric (graduated) flask and made up to 250.0 cm³. 25.0 cm³ portions of this solution were placed in a conical flask and 20 cm³ of dilute sulfuric acid were added. This mixture was titrated against a standard solution of potassium manganate(VII) containing 0.0204 mol dm⁻³.

The student's results are shown in the table below.

Titration	1	2	3	4
Final/cm³	23.8	23.35	23.45	25.60
Initial/cm³	0.5	0.45	1.75	2.70

(a) **(i)** Calculate the value of x in the salt. Show all your working. Include in your answer an ionic equation for the reaction.

_____

_____

_____

_____

_____

_____

_____

_____

_____

_____

_____ 8 marks

(ii) Explain why the student has given the results for titration 1 to 1 decimal place.

_____

_____ 2 marks

(iii) Calculate the percentage error in using the burette in this experiment.

_____

_____ 2 marks

(iv) Explain why the M_r of the salt need not be known to a high accuracy in order to be confident of the value of x in the formula.

_____

_____ 1 mark

(b) **(i)** Explain why sulfuric acid was added to the iron(II) solution before titration.

_____

_____ 2 marks

(ii) Suggest what piece of laboratory equipment is best suited for measuring this sulfuric acid solution. Explain your answer.

_____

_____ 2 marks

Total marks: 17

18 Some electrode potentials are shown in the table below. Use these to help you answer the questions that follow.

Half-equation	$E^{\ominus}$/V
$VO_2^+(aq) + 2H^+(aq) + e^- \rightarrow VO^{2+}(aq) + H_2O(l)$	+1.00
$VO^{2+}(aq) + 2H^+(aq) + e^- \rightarrow V^{3+}(aq) + H_2O(l)$	+0.34
$V^{3+}(aq) + e^- \rightarrow V^{2+}(aq)$	−0.26
$Zn^{2+}(aq) + 2e^- \rightarrow Zn(s)$	−0.76
$V^{2+}(aq) + 2e^- \rightarrow V(s)$	−1.13

(a) Explain why zinc in aqueous acid conditions reduces V^{3+} ions to V^{2+} ions but cannot reduce V^{2+} ions to metallic vanadium.

_____

_____ 2 marks

(b) Zinc reacts with vanadate(V) ions in acid conditions to form vanadium(II) ions.

   (i) Write an overall equation for the reaction.

_____ 1 mark

   (ii) Give the colours you would observe during the reaction and for each colour identify the oxidation state of the vanadium species responsible for the colour.

_____

_____

_____

_____ 4 marks

(c) Give the standard representation of the cell which would give a standard EMF of + 1.10 V. Use platinum electrodes where necessary.

_____

_____ 2 marks

_____ Total marks: 9

19 Give final observation(s) and write (an) ionic equation(s) for each of the following reactions

(a) $CuSO_4(aq)$ and an excess of concentrated $HCl(aq)$

   _Observation_____

   _Equation_____ 3 marks

(b) $FeSO_4(aq)$ and $BaCl_2(aq)$

   _Observation_____

   _Equation_____ 2 marks

(c) $FeCl_3$(aq) and dilute NH_3(aq)

Observation _____

Equation _____ 3 marks

(d) $Al_2(SO_4)_3$(aq) and NaOH(aq) until in excess

Observations _____

_____

Equations _____

_____ 4 marks

(e) AgBr(s) and concentrated NH_3(aq)

Observation _____

Equation _____ 2 marks

Total marks: 14

20 (a) Many transition-metal ions act as homogenous catalysts in chemical reactions. Explain what is meant by the term *homogeneous* in this context.

_____ 1 mark

(b) The reaction between peroxodisulfate ions ($S_2O_8{}^{2-}$ ions) and iodide ions is catalysed by transition-metal ions. Explain how this reaction is catalysed by a transition-metal ion writing equations to illustrate your answer.

_____

_____

_____

_____

_____

_____

_____

_____

_____

_____

_____

_____

_____

_____

_____

_____

_____

_____ 7 marks

Total marks: 8

Multiple choice questions

1 Which of the following statements about transition-metal ions is correct?

A The colour change at the endpoint of a manganate(VII)–iron(II) titration when manganate(VII) is added from a burette is pink to colourless.

B Manganate(VII) ions act as a homogeneous catalyst in the reaction between manganate(VII) ions and ethanedioate ions.

C The aerial oxidation of iron(II) ions is more favourable in acidic conditions.

D A linear complex ion is formed when an excess of dilute aqueous ammonia is added to solid silver chloride.

2 Manganate(VII) ions oxidise hydrogen peroxide in acidic solution. The half-equation for the oxidation reaction is:

$$H_2O_2 \rightarrow O_2 + 2H^+ + 2e^-$$

$25.0 \ cm^3$ of a solution of hydrogen peroxide are oxidised exactly by $26.4 \ cm^3$ of $0.0200 \ mol \ dm^{-3}$ potassium manganate(VII).

The concentration, in $mol \ dm^{-3}$, of the hydrogen peroxide is

A 0.106

B 0.0528

C 0.0211

D 0.00837

3 Which of the following would give a solid residue when an excess of dilute aqueous ammonia is added?

A $CuSO_4(aq)$

B $Ba(NO_3)_2(aq)$

C $Al_2(SO_4)_3(aq)$

D $AgCl(s)$

4 A buffer solution is made by adding 0.200 mol of sodium propanoate to $1 \ dm^3$ of a $0.400 \ mol \ dm^{-3}$ solution of propanoic acid ($K_a = 1.51 \times 10^{-5} \ mol \ dm^{-3}$). The pH of the solution formed is:

A 2.61

B 4.52

C 4.82

D 5.12

Answers

Question	Answer		Marks
1 (a)	*Catalyst* vanadium(V) oxide	(1)	
	Meaning of heterogeneous catalyst and reactants in different phases	(1)	2
1 (b)	to prevent poisoning of the catalyst	(1)	1
1 (c)	*Observations* burns with blue flame	(1)	
	pungent gas formed	(1)	
	Equation $S + O_2 \rightarrow SO_2$	(1)	3
1 (d) (i)	*Amount of oxygen* 0.085 mol	(1)	
	Amount of sulfur trioxide 1.09 mol	(1)	2
1 (d) (ii)	*Mole fraction of sulfur dioxide* 0.152		
	Mole fraction of oxygen 0.0614		
	Mole fraction of sulfur trioxide 0.787	(1)	1
1 (d) (iii)	partial pressure of sulfur dioxide = $110\,000 \times 0.152 = 16720$ Pa		
	partial pressure of oxygen = $110\,000 \times 0.0614 = 6754$ Pa		
	partial pressure of sulfur trioxide = $110\,000 \times 0.787 = 86570$ Pa (1) for three partial pressures		
	$K_p = \dfrac{pSO_3{}^2}{pSO_2{}^2 \times pO_2}$	(1)	
	$= (86\,570)^2/\{(16\,720)^2(6754)\}$	(1)	
	$= 3.96918 \times 10^{-3}$		
	$= 4.0 \times 10^{-3}$ (2 s.f.)	(1)	
	units: Pa^{-1}	(1)	5
1 (e)	K_p is smaller at T_2 than T_1, so equilibrium has moved to the left	(1)	
	ΔH is −ve, so the reaction is exothermic to the right	(1)	
	since equilibrium has moved towards the endothermic side, T_2 is higher than T_1	(1)	3
			Total 17
2 (a)	enthalpy change when one mole of gaseous atoms	(1)	
	are formed from the element in its standard state	(1)	2
2 (b) (i)			
	Atomisation of oxygen step can be first step		4

Question	Answer		Marks
2 (b) (ii)	$\Delta_f H + \Delta_{\text{lattice dissociation}} H = \Delta_{\text{atomisation}} H(\text{Ca}) + \Delta_{\text{first IE}} H(\text{Ca}) + \Delta_{\text{second IE}} H(\text{Ca}) + \Delta_{\text{atomisation}} H(\text{O})$		
	$+ \Delta_{\text{first EA}} H(\text{O}) + \Delta_{\text{second EA}} H(\text{O})$	(1)	
	$-635 + 3646 = +178 + 732 + 1150 + 249 + (-142) + \Delta_{\text{second EA}} H(\text{O})$	(1)	
	$\Delta_{\text{second EA}} H(\text{O}) = +844 \text{ kJ mol}^{-1}$	(1)	3
2 (b) (iii)	electron gained is attracted to the nucleus	(1)	1
2 (b) (iv)	electron is being added to a negative ion	(1)	
	there is repulsion between the electron and the negative ion	(1)	2
			Total 12
3 (a)	ionic lattice is formed from perfectly spherical ions which can be represented as point charges	(1)	
	with 100% ionic bonding and no covalent character	(1)	2
3 (b) (i)	for NaCl, lattice formation enthalpy is more exothermic than for KCl because there is stronger attraction between Na^+ and Cl^- ions than between K^+ and Cl^- ions	(1)	
	this is because Na^+ ions are smaller than K^+ ions	(1)	2
3 (b) (ii)	100% ionic bonding with no covalent character	(1)	
	because there is good agreement between theoretical and experimental values	(1)	2
3 (b) (iii)	some covalent character because poor agreement	(1)	
	experimental value more exothermic than expected because covalent bonding is stronger than ionic bonding	(1)	2
			Total 8
4 (a) (i)	enthalpy change when one mole of gaseous ions	(1)	
	are hydrated	(1)	2
4 (a) (ii)	$\Delta_{\text{solution}} H = \Delta_{\text{lattice dissociation}} H + \Sigma \Delta_{\text{hydration}} H(\text{ions})$	(1)	
	$-155 = +2493 + (-1926) + 2 \times \Delta_{\text{hydration}} H(\text{Cl}^-)$	(1)	
	$2 \times \Delta_{\text{hydration}} H(\text{Cl}^-) = -722 \text{ kJ mol}^{-1}$		
	$\Delta_{\text{hydration}} H(\text{Cl}^-) = -361 \text{ kJ mol}^{-1}$	(1)	3
4 (b)	Hydration enthalpy of the fluoride ion is much more exothermic than that of the bromide ion	(1)	
	there are stronger attractions between fluoride ions and polar water molecules than there are between bromide ions and water	(1)	
	because fluoride ions are smaller than bromide ions	(1)	3
			Total 8
5 (a)	x-axis labelled time/mins and y-axis labelled temperature/°C with values that use more than half of each axis	(1)	
	all points plotted correctly	(1)	
	two straight lines drawn	(1)	
	ignoring points at 1 min/18.1 °C, 5 min/13.5 °C and 9 min/12.4 °C	(1)	
	Temperature change at the fourth minute 6.7 K	(1)	5

Question	Answer		Marks
5 (b)	Heat change = $mc\,\delta T$	(1)	
	$\quad\quad\quad = 25 \times 4.2 \times 6.7$ J	(1)	
	$\quad\quad\quad = 703.5$ J		
	Moles of sodium hydrogencarbonate = mass/M_r = 2.14/84.0 = 0.0255 mol		
	moles of hydrochloric acid = $c \times v$/1000 = 1.10 $\times$ 25.0 /1000 = 0.0275 mol	(1)	
	so hydrochloric acid is in excess and enthalpy change depends on sodium hydrogencarbonate		
	$\Delta_{combustion}H$ = heat change/moles = +703.5/0.0255 J mol^{-1}	(1)	
	$\quad\quad = +27588$ J mol^{-1}		
	$\quad\quad = + 28$ kJ mol^{-1} (2 sig. figs.)	(1)	5
5 (c) (i)	percentage uncertainty in measuring temperature = 0.2/6.7 $\times$ 100 = 3.0%	(1)	
	percentage uncertainty in measuring mass = 0.01/2.41 $\times$ 100 = 0.4%	(1)	
	percentage uncertainty in measuring volume = 0.1/25.0 $\times$ 100 = 0.4%	(1)	
	so the student was correct	(1)	
	use a greater mass of sodium hydrogencarbonate **and** a more concentrated solution of hydrochloric acid so there is a greater temperature change	(1)	5
5 (c) (ii)	the hydrochloric acid is in excess	(1)	
	so does not control the temperature change	(1)	2
5 (d) (i)	difference = 28.5 – 28.0 = 0.5		
	percentage difference = 0.5/28.5 $\times$ 100 = 1.8%	(1)	1
5 (d) (ii)	reaction is endothermic so ΔH is positive	(1)	
	for a reaction to be spontaneous $\Delta G \leq 0$	(1)	
	$\Delta G = \Delta H - T\Delta S$	(1)	
	ΔS in this reaction is positive because gas is formed	(1)	
	at this temperature the $T\Delta S$ term outweighs the ΔH term and makes ΔG negative	(1)	5
			Total 23
6 (a)	$\Delta G = \Delta H - T\Delta S$	(1)	1
6 (b)	$\Delta G \leq 0$	(1)	1
6 (c) (i)	*Sign of* $\Delta H^{\ominus}$ negative	(1)	
	Sign of $\Delta S^{\ominus}$ negative	(1)	2
6 (c) (ii)	$- T\Delta S$ term is positive (because ΔS is negative and T is positive)	(1)	
	when $T > 273$, $T\Delta S$ outweighs ΔH term and makes ΔG positive	(1)	2
6 (d) (i)	$\Delta_{reaction}H^{\ominus} = \Sigma \Delta_f H^{\ominus}$(products) – $\Sigma \Delta_f H^{\ominus}$(reactants)	(1)	
	$\therefore +18 = [(3 \times \Delta H_f(KClO_4) + (-436)] - [4 \times (-391)]$	(1)	
	$\therefore 3 \times \Delta_f H\ (KClO_4) = -1110$		
	$\therefore \Delta_f H(KClO_4) = -370$ kJ mol^{-1}	(1)	3
6 (d) (ii)	$\Delta_{reaction}S = \Sigma\ S$(products) – $\Sigma\ S$(reactants)	(1)	
	$\quad = [(3 \times 134) + (83)] - [4 \times (112)]$	(1)	
	$\quad = +37$ J K^{-1} mol^{-1}	(1)	3
6 (d) (iii)	$\Delta G = \Delta H - T\Delta S$	(1)	
	$\quad = 18 - (298 \times 37)/1000$		
	$\quad = + 7.0$	(1)	
	$\quad\quad$ kJ mol^{-1}	(1)	3

Question	Answer		Marks
6 (d) (iv)	$T = \Delta H / \Delta S$	(1)	
	$= 18 \times 1000/37$	(1)	
	$= 486$ K (or 213 °C) (units essential)	(1)	3
			Total 18
7 (a)	$\Delta G = \Delta H - T\Delta S$	(1)	1
7 (b)	*Value of the slope* 44	(1)	
	Units J K^{-1} mol^{-1}	(1)	
	Symbol ΔS	(1)	3
7 (c)	ΔG is negative	(1)	
	formation of water is feasible below this temperature	(1)	2
7(d)	**This answer is marked using levels of response.**		
	Level 3: 5–6 marks		
	All parts are covered and the explanation of each part is generally correct and virtually complete.		
	Answer communicates the whole process coherently and shows a logical progression from part 1 and part 2 to overall explanation.		
	Level 2: 3–4 marks		
	All parts are covered but the explanation of each part may be incomplete OR two parts are covered and the explanations are virtually complete.		
	Answer is mainly coherent and shows a progression. Some statements may be out of order and incomplete.		
	Level 1: 1–2 marks		
	Two parts are covered but the explanation of each part may be incomplete and contain inaccuracies OR only one part is covered but the explanation is mainly correct and is virtually complete.		
	Answer includes some isolated statements but there is no attempt to present them in a logical order or show confused reasoning.		
	Level 0: 0 marks		
	equation and enthalpy change:		
	at 373 K and above: $CH_3OH(g) + 1\frac{1}{2}O_2(g) \rightarrow CO_2(g) + 2H_2O(g)$		
	ΔH is large and negative as combustion is an exothermic reaction		
	entropy change		
	at 373 K and above: 2½ mol of gas becomes 3 mol of gas		
	increase in disorder		
	so ΔS is positive		
	discussion of ΔG		
	$\Delta G = \Delta H - T\Delta S$		
	since ΔH is negative and $-T\Delta S$ is negative at all temperatures, ΔG is negative		
	if ΔG is negative, reaction is spontaneous	6	
			Total 12
8 (a)	gives electrons to another species	(1)	1
8 (b) (i)	*Standard reference electrode* standard hydrogen electrode	(1)	
	Electrode potential 0.0 V	(1)	
	Explanation by convention	(1)	3

Question	Answer		Marks
8 (b) (ii)	*Condition 1* pressure 100 kPa	(1)	
	Condition 2 concentration of H^+ ions 1 mol dm^{-3}	(1)	
	Condition 3 298 K	(1)	3
8 (b) (iii)	*Function* allow transfer of ions and complete the circuit	(1)	
	Contents potassium nitrate (other inert soluble ionic salts may also be used)	(1)	2
8 (c)	series of reduction half–equations	(1)	
	arranged in order of electrode potentials	(1)	2
8 (d) (i)	$Zn^{2+} + 2e^- \rightarrow Zn$	(1)	1
8 (d) (ii)	*Cell EMF* 0.32 V	(1)	1
8 (d) (iii)	*Species reduced* Fe^{2+}	(1)	
	Explanation Fe^{2+}/Fe $E^{\ominus}$ is less negative than Zn^{2+}/Zn $E^{\ominus}$	(1)	2
			Total 15
9 (a) (i)	not rechargeable	(1)	1
9 (a) (ii)	*Benefit* cheap	(1)	
	Risk disposal hazard to environment	(1)	2
9 (a) (iii)	*Oxidation reaction* $Li \rightarrow Li^+ + e^-$	(1)	
	Reduction reaction $MnO_2 \rightarrow MnO_2^- + e^-$	(1)	
	Oxidation state of Mn in $LiMnO_2$ +3/III	(1)	3
9 (b) (i)	*Oxidising agent* MnO_2	(1)	
	Oxidation state of element before reaction +4/IV	(1)	
	Oxidation state of element after reaction +3/III	(1)	3
9 (b) (ii)	the reaction is not reversible	(1)	1
9 (c) (i)	the cell is rechargeable	(1)	1
9 (c) (ii)	$PbO_2(s) + 3H^+(aq) + HSO_4^-(aq) + 2e^- \rightarrow PbSO_4(s) + 2H_2O(l)$ $E^{\ominus} = +1.68$ V		
	$PbO_2(s) + Pb(s) + 2H^+(aq) + 2HSO_4^-(aq) \rightleftharpoons 2PbSO_4(s) + 2H_2O(l)$		
	$EMF = +2.04$ V		
	two equations re-arranged		
	subtract		
	Half-equation $PbSO_4 + H^+ + 2e^- \rightarrow Pb + HSO_4^-$	(1)	
	Value of $E^{\ominus}$= 1.68 – 2.04 V		
	= –0.36 V	(1)	2
9 (d) (i)	*Half-equation* $H_2 + 2OH^- \rightarrow 2H_2O + 2e^-$	(1)	1
9 (d) (ii)	*Half-equation* $O_2 + 2H_2O + 4e^- \rightarrow 4OH^-$	(1)	1
9 (d) (iii)	$4OH^- + 2H_2 \rightarrow 4H_2O + 4e^-$ 0.40 V		
	$O_2 + 2H_2O + 4e^- \rightarrow 4OH^-$ 0.83 V		
	two equations rearranged		
	add and cancel		
	Overall equation $2H_2 + O_2 \rightarrow 2H_2O$	(1)	
	Cell EMF 1.23 V	(1)	2
9 (d) (iv)	*Overall cell equation* $2H_2 + O_2 \rightarrow 2H_2O$	(1)	
	$E^{\ominus}(O_2/H_2O)$ 1.23 V	(1)	2

Question	Answer		Marks
9 (d) (v)	Horizontal line	(1)	
	Explanation voltage is constant as long as fuel and oxygen are constantly supplied	(1)	2
			Total 21
10 (a) (i)	*Brønsted–Lowry acid* H^+ donor	(1)	
	Strong acid acid fully dissociated into ions	(1)	2
10 (a) (ii)	$pH = -\log[H^+] = -\log(1.19) = -0.08$	(1)	1
10 (b) (i)	*Definition of K_w:* $K_w = [H^+][OH^-]$	(1)	
	Value at 298: 1.00×10^{-14}	(1)	
	Units of K_w $mol^2\,dm^{-6}$	(1)	3
10 (b) (ii)	$H_2O \rightleftharpoons H^+ + OH^-$ (equilibrium sign required)	(1)	1
10 (b) (iii)	$K_w = [H^+][OH^-] = [H^+]^2$	(1)	
	$[H^+] = \sqrt{5.13 \times 10^{-13}}$	(1)	
	$[H^+] = 7.162 \times 10^{-13}\,mol\,dm^{-3}$		
	$pH = 6.14$	(1)	3
10 (b) (iv)	in pure water, $[H^+] = [OH^-]$	(1)	1
10 (b) (v)	*Sign of enthalpy change* positive	(1)	
	Explanation as temperature increases K_w increases so equilibrium moves to right		
	as temperature increases, equilibrium moves in endothermic direction	(1)	2
10 (c) (i)	$[H^+] = K_w/[OH^-] = 5.48 \times 10^{-14}/0.131$	(1)	
	$[H^+] = 4.183 \times 10^{-13}$		
	$pH = 12.38$	(1)	2
10 (c) (ii)	moles of NaOH $= c \times v/1000 = 0.131 \times 25.0/1000 = 3.275 \times 10^{-3}$		
	moles of $H_2SO_4 = c \times v/1000 = 0.140 \times 12.5/1000 = 1.75 \times 10^{-3}$ (1) for both		
	$\therefore$ moles of $OH^- = 3.275 \times 10^{-3}$ and moles of $H^+ = 2 \times 1.75 \times 10^{-3} = 3.50 \times 10^{-3}$	(1)	
	$\therefore$ moles of H^+ in excess $= 3.50 \times 10^{-3} - 3.275 \times 10^{-3} = 2.25 \times 10^{-4}$	(1)	
	$\therefore [H^+] = n/v = 2.25 \times 10^{-4}/(37.5/1000) = 6.00 \times 10^{-3}$	(1)	
	$\therefore pH = 2.22$	(1)	5
			Total 20
11 (a) (i)	*x*-axis labelled volume (of sodium hydroxide)/cm^3 and *y*-axis pH, with values that use more than half of each axis	(1)	
	all points plotted correctly	(1)	
	smooth curve drawn	(1)	
	ignoring points at 6 cm^3/4.2 and 16 cm^3/5.1	(1)	4
11 (a) (ii)	*Volume of sodium hydroxide* 22.0 cm^3	(1)	
	Concentration of butanoic acid $n(NaOH) = c \times v/1000 = 0.100 \times 22.0/1000 = 0.00220\,mol$		
	1:1 so n(butanoic acid) $= 0.00220$ mol	(1)	
	$\therefore c$(butanoic acid) $= 0.00220/(25.0/1000) = 0.0880\,mol\,dm^{-3}$	(1)	3

Question	Answer		Marks
11 (a) (iii)	$K_a = 10^{-pKa} = 1.479 \times 10^{-5}$ (mol dm^{-3})	(1)	
	$[H^+] = \sqrt{(K_a \times [\text{Butanoic acid}])} = \sqrt{(1.479 \times 10^{-5} \times 0.0880)} = 1.1409 \times 10^{-3}$ (mol dm^{-3})	(1)	
	pH = 2.94	(1)	3
11 (b)	pH = 4.8	(1)	
	at half-equivalence, pH = pK_a = 4.8	(1)	
	$K_a = 10^{-4.8} = 1.6 \times 10^{-5}$ (mol dm^{-3})	(1)	3
11 (c) (i)	*Identical feature 1* volume of sodium hydroxide at equivalence is same	(1)	
	Identical feature 2 curve shape is same after equivalence	(1)	2
11 (c) (ii)	*Different features* pH with no added alkali is different	(1)	
	shape of curve before equivalence is different	(1)	
	pH at equivalence is different	(1)	
	any 2		2
11 (d) (i)	cresolphthalein	(1)	1
11 (d) (ii)	phenol red	(1)	1
11 (e)	pH meter is used to measure buffers of known concentration	(1)	
	calibration curve is plotted of measured pH against actual/buffer pH	(1)	
	the measured pH of pH meters drifts over time	(1)	3
			Total 22
12 (a)	A solution which resists change in pH when acid or alkali is added	(1)	
	or on dilution with water	(1)	2
12 (b) (i)	*Acidic buffer* one with a pH of less than 7	(1)	1
12 (b) (ii)	*Reagents* a solution of a named weak acid and a named salt of the acid or a weak acid and a named strong base e.g. ethanoic acid and sodium ethanoate or ethanoic acid and sodium hydroxide etc.	(1)	
	Equation e.g. $CH_3COOH \rightleftharpoons CH_3COO^- + H^+$	(1)	2
12 (b) (iii)	when H^+ is added it reacts with CH_3COO^- and equilibrium position moves to the left	(1)	
	$[H^+]$ does not increase by much so pH only falls by a small amount	(1)	
	when OH^- is added it reacts with H^+ ions and equilibrium position moves to the right to replace H^+ ions	(1)	
	$[H^+]$ does not decrease by much so pH only rises by a small amount	(1)	4
12 (c)	$K_a = [H^+][X^-]/[HX]$ $\therefore [H^+] = K_a \times [HX]/[A^-]$	(1)	
	$K_a = 10^{-pKa} = 1.00 \times 10^{-5}$	(1)	
	original moles of HX = 20.0/1000 $\times$ 0.24 = 0.0048 mol		
	moles of OH^- added = 10.0/1000 $\times$ 0.20 = 0.0020 mol	(1) for both	
	moles of X^- formed = 0.0020 mol		
	moles of HX remaining = 0.0048 – 0.0020 = 0.0028 mol	(1) for both	
	let volume of mixture be v cm^3		
	$[H^+] = 1.00 \times 10^{-5} \times (0.0028/v)/0.0020/v)$ (and v's cancel)	(1)	
	$[H^+] = 1.40 \times 10^{-5}$		
	pH = 4.85	(1)	6
			Total 15

Question	Answer		Marks
13(a)	**This answer is marked using levels of response.**		
	Level 3: 5–6 marks All parts are covered and the explanation of each part is generally correct and virtually complete. Answer communicates the whole process coherently and shows a logical progression from part 1 and part 2 to overall explanation.		
	Level 2: 3–4 marks All parts are covered but the explanation of each part may be incomplete OR two parts are covered and the explanations are virtually complete.. Answer is mainly coherent and shows a progression. Some statements may be out of order and incomplete.		
	Level 1: 1–2 marks Two parts are covered but the explanation of each part may be incomplete and contain inaccuracies OR only one part is covered but the explanation is mainly correct and is virtually complete. Answer includes some isolated statements but there is no attempt to present them in a logical order or show confused reasoning.		
	Level 0: 0 marks		

	bonding:
	  magnesium oxide is ionic
	  phosphorus(V) oxide is covalent
	structures:
	  magnesium oxide is a giant ionic lattice
	  phosphorus(V) oxide is molecular
	link of structure and bonding to melting point:
	  MgO: strong electrostatic attraction between opposite charged ions
	    needs high energy to overcome
	  P_4O_{10}: weak van der Waals' forces between molecules
	    needs much less energy to overcome
			6

13 (b)	*Observation with magnesium oxide* no visible change/sparingly soluble	(1)	
	Equation with magnesium oxide $MgO + H_2O \rightarrow Mg(OH)_2$	(1)	
	pH of mixture answer within range 8–10	(1)	
	Observation with phosphorus(V) oxide violent reaction and dissolves	(1)	
	Equation with phosphorus(V) oxide $P_4O_{10} + 6H_2O \rightarrow 4H_3PO_4$	(1)	
	pH of mixture answer within the range −1 to +1	(1)	6
13 (c) (i)	*Equation* $4Al + 3O_2 \rightarrow 2Al_2O_3$ (or multiples/fractions)	(1)	
	Explanation aluminium forms a surface layer of oxide which is non-porous	(1)	2
13 (c) (ii)	aluminium oxide has ionic lattice with high lattice enthalpy/very strong electrostatic attraction between the opposite charged ions	(1)	
	water is not able to break the forces in the lattice/hydration enthalpy not sufficiently exothermic to outweigh lattice enthalpy	(1)	
	silicon dioxide is macromolecular with strong covalent bonds	(1)	
	water cannot break the covalent bonds	(1)	4

Question	Answer		Marks
13 (c) (iii)	Equation for aluminium oxide $Al_2O_3 + 6OH^- + 3H_2O \rightarrow 2[Al(OH)_6]^{3-}$		
	or $Al_2O_3 + 2OH^- + 7H_2O \rightarrow 2[Al(OH)_4(H_2O)_2]^-$	(1)	
	Equation for silicon dioxide $SiO_2 + 2OH^- \rightarrow SiO_3^{2-} + H_2O$	(1)	2
			Total 20
14 (a)	Oxidation state of chromium +3	(1)	
	Co-ordination number of chromium 6	(1)	2
14 (b) (i)	Electronic configuration: [Ar] $3d^5$ in each	(1)	
	Coordination number: 6 in each	(1)	2
14 (b) (ii)	visible light of a certain colour/frequency/wavelength is absorbed	(1)	
	electrons are promoted from a lower energy d-orbital to a higher energy d-orbital	(1)	
	the light not absorbed is transmitted and is the colour we see	(1)	3
14 (b) (iii)	$E = hf$ (or $h\nu$), so $f = E/h = 4.139 \times 10^{-19}/6.62607 \times 10^{-34}$	(1)	
	$= 6.246338 \times 10^{14}$		
	$= 6.246 \times 10^{14}\,s^{-1}$ (answer to 4 sig. fig.)	(1)	2
14 (c) (i)	a ligand which donates lone pairs	(1)	
	from two different atoms in the ligand	(1)	2
14 (c) (ii)	ligand substitution	(1)	1
14 (c) (iii)			
	Fe 6 coordinate with O atoms	(1)	
	structure of ligands shown correctly	(1)	
	isomers are mirror images of each other	(1)	3
			Total 15
15 (a)	$TiCl_4 + 2H_2O \rightarrow TiO_2 + 4HCl$	(1)	1
15 (b) (i)	Ti(IV) [Ar] $3d^0$	(1)	
	Ti(III) [Ar] $3d^1$	(1)	2
15 (b) (ii)	Ti(III) has an incomplete d-subshell/has partially filled d orbitals	(1)	
	Ti(IV) has an empty d-subshell/has no electrons in d orbitals	(1)	2
15 (b) (iii)	use colorimeter/visible spectrometer	(1)	
	make several solutions of known concentration of Ti(III) ions	(1)	
	measure their absorbances	(1)	
	plot calibration curve of absorbance against concentration	(1)	
	measure absorbance of unknown solution	(1)	
	read off concentration from calibration curve	(1)	6

Question	Answer	Marks
15 (c)	$Ti^{3+} \rightarrow Ti^{4+} + e^- \qquad \times 5$ $MnO_4^- + 8H^+ + 5e^- \rightarrow Mn^{2+} + 4H_2O \qquad \times 1 \qquad$ add $MnO_4^- + 8H^+ + 5Ti^{3+} \rightarrow Mn^{2+} + 4H_2O + 5Ti^{4+} \qquad$ (1) for species (1) for balance	2
15 (d)	*Observations* coloured precipitate $\qquad$ (1) effervescence/bubbling/fizzing $\qquad$ (1) *Equation* $2Ti(H_2O)6^{3+} + 3CO_3^{2-} \rightarrow 2Ti(OH)_3(H_2O)_3 + 3CO_2 + 3H_2O \qquad$ (1) for species $\qquad$ (1) for balance	4
		Total 17
16 (a) (i)	*Equation* $CuSO_4 + 6H_2O \rightarrow [Cu(H_2O)_6]^{2+} + SO_4^{2-} \qquad$ (1)	1
16 (a) (ii)	*Equation for formation of pale blue precipitate* $[Cu(H_2O)_6]^{2+} + 2NH_3 \rightarrow [Cu(OH)_2(H_2O)_4] + 2NH_4^+ \qquad$ (1) for product copper complex, $\qquad$ (1) for balance *Equation for conversion of blue precipitate into deep blue solution* $[Cu(OH)_2(H_2O)_4] + 4NH_3 \rightarrow [Cu(NH_3)_4(H_2O)_2]^{2+} + 2H_2O + 2OH^- \qquad$ (1) for product copper complex, (1) for balance	4
16 (a) (iii)	*Equation* $[Cu(NH_3)_4(H_2O)_2]^{2+} + 2H_2NCH_2CH_2NH_2 \rightarrow [Cu(H_2NCH_2CH_2NH_2)_2(H_2O)_2]^{2+} + 4NH_3$ $\qquad$ (1) for product copper complex, (1) for balance *Explanation* 3 moles of reactant particles becomes 5 moles of product particles/ there are more moles of product particles than reactant particles $\qquad$ (1) ΔS/entropy change for the reaction is positive $\qquad$ (1) and since ΔH is small, $\Delta G \leq 0$ and the reaction is spontaneous $\qquad$ (1)	5
16 (b) (i)	*Use of cisplatin* chemotherapy/cancer treatment $\qquad$ (1) *Risk of its use* can kill healthy cells as well as cancer cells $\qquad$ (1)	2
16 (b) (ii)	 (1) for co-ordinate bonds from N and Cl, (1) for *trans/E*-geometry	2
16 (c) (i)	*Formula* $[Ag(NH_3)_2]^+ \qquad$ (1) *Bond angle* 180° $\qquad$ (1)	2
16 (c) (ii)	aldehyde or CHO $\qquad$ (1)	1
16 (c) (iii)	$RCHO + 3OH^- + 2[Ag(NH_3)_2]^+ \rightarrow RCOO^- + 2Ag + 4NH_3 + 2H_2O$ or $RCHO + 2H_2O + 2[Ag(NH_3)_2]^+ \rightarrow RCOOH + 2Ag + 2NH_3 + 2NH_4^+ \qquad$ (1) for carboxylic acid/ carboxylate anion, (1) for balance	2
		Total 19
17 (a) (i)		

Titration	1	2	3	4
Final/cm³	23.8	23.35	23.45	25.60
Initial/cm³	0.5	0.45	1.75	2.70
Titre	**23.3**	**22.90**	**22.70**	**22.90**

Question	Answer		Marks
	Average titre 22.90 cm^3 (using concordant titres 2 and 4)	(1)	
	Moles of MnO_4^- = $c \times v/1000$ = 0.0204 $\times$ 22.9/1000 = 4.6716 $\times 10^{-4}$ mol	(1)	
	$MnO_4^- + 8H^+ + 5Fe^{2+} \rightarrow Mn^{2+} + 4H_2O + 5Fe^{3+}$	(1)	
	mol of Fe(II) in 25.0 cm^3 = $5 \times MnO_4^-$ = 2.3358×10^{-3}	(1)	
	mol of Fe(II) in 250.0 cm^3 = $5 \times MnO_4^-$ = 2.3358×10^{-2}	(1)	
	M_r of salt = mass/moles = 6.45/2.3358×10^{-2} = 276.1	(1)	
	M_r of $FeSO_4$ = 151.9		
	$\therefore M_r$ of xH_2O = 276.1 − 151.9 = 124.2	(1)	
	$\therefore$ number of waters = 124.2/18.0 = 6.9		
	$\therefore x = 7$	(1)	8
17 (a) (ii)	rough titration	(1)	
	so less precision is warranted in values	(1)	2
17 (a) (iii)	error in two burette readings = 2 $\times$ 0.05 cm^3 = 0.1 cm^3	(1)	
	percentage error = 0.1/22.9 $\times$ 100 = 0.44 %	(1)	2
17 (a) (iv)	x must be a whole number so as long as the number of moles of water is close to a whole number we can be confident the answer is correct	(1)	1
17 (b) (i)	H$^+$ ions take part in the reaction	(1)	
	this ensures H$^+$ ions are in excess	(1)	2
17 (b) (ii)	measuring cylinder	(1)	
	since H$^+$ ions are in excess, the volume does not need to be accurate or precisely known	(1)	2
			Total 17
18 (a)	$E^{\ominus}$ for Zn^{2+}/Zn is more negative than $E^{\ominus}$ for V^{3+}/V^{2+}	(1)	
	$E^{\ominus}$ for Zn^{2+}/Zn is more positive than $E^{\ominus}$ for V^{2+}/V	(1)	
	or		
	$E^{\ominus}$ for V^{3+}/V^{2+} is more positive than $E^{\ominus}$ for Zn^{2+}/Zn	(1)	
	$E^{\ominus}$ for V^{2+}/V is more negative than $E^{\ominus}$ for Zn^{2+}/Zn	(1)	2
18 (b) (i)	$3Zn + 2VO_2^+ + 8H^+ \rightarrow 3Zn^{2+} + 2V^{2+} + 4H_2O$	(1)	1
18 (b) (ii)	vanadium(V) – yellow	(1)	
	vanadium(IV) – blue	(1)	
	vanadium(III) – green	(1)	
	vanadium(II) – violet	(1)	4
18 (c)	$Zn|Zn^{2+}||VO^{2+},H^+,V^{3+}|Pt$ (1) for Zn and V species in correct order, (1) for phase boundaries, salt bridge and Pt		2
			Total 9
19 (a)	*Observation* (yellow/)green solution	(1)	
	Equation $[Cu(H_2O)_6]^{2+} + 4Cl^- \rightarrow [CuCl_4]^{2-} + 6H_2O$ (1) for copper complexes, (1) for balance		3
19 (b)	*Observation* white precipitate	(1)	
	Equation $Ba^{2+} + SO4^{2-} \rightarrow BaSO_4$	(1)	2
19 (c)	*Observation* brown precipitate	(1)	
	Equation $[Fe(H_2O)_6]^{3+} + 3NH_3 \rightarrow Fe(OH)_3(H_2O)_3 + 3NH_4^+$ (1) for iron complexes, (1) for balance		3

Question	Answer		Marks
19 (d)	*Observations* white precipitate	(1)	
	dissolves in excess to give colourless solution	(1)	
	Equations $[Al(H_2O)_6]^{3+} + 3OH^- \rightarrow Al(OH)_3(H_2O)_3 + 3H_2O$	(1)	
	either $Al(OH)_3(H_2O)_3 + 3OH^- \rightarrow [Al(OH)_6]^{3-} + 3H_2O$	(1)	
	or $Al(OH)_3(H_2O)_3 + OH^- \rightarrow [Al(OH)_4(H_2O)_2]^- + H_2O$	(1)	4
19 (e)	*Observation* colourless solution	(1)	
	Equation $AgBr + 2NH_3 \rightarrow [Ag(NH_3)_2]^+$	(1)	2
			Total 14
20 (a)	catalyst and reactants are in the same phase	(1)	1
20 (b)	Overall reaction: $S_2O_8^{2-} + 2I^- \rightarrow 2SO_4^{2-} + I_2$	(1)	
	Step 1: $S_2O_8^{2-} + 2Fe^{2+} \rightarrow 2SO_4^{2-} + 2Fe^{3+}$	(1)	
	Step 2: $2Fe^{3+} + 2I^- \rightarrow 2Fe^{2+} + I_2$	(1)	
	reaction is slow in the absence of a catalyst because two anions are reacting together		
	so repel	(1)	
	catalyst: Fe^{2+} or Fe^{3+}	(1)	
	iron ions provide an alternative route with a lower E_a	(1)	
	because opposite charged ions are reacting together so reaction is speeded up and Fe^{2+} (or Fe^{3+}) is re-formed at the end of the reaction	(1)	7
			Total 8

Multiple choice questions

1 D	
2 B	
3 C	
4 B	

Glossary

$\Delta H^{\ominus}$ (298 K)	the standard enthalpy change at 298 K
absolute entropy	see *standard entropy*
acid–base equilibrium	the equilibrium transfer of a proton from an acid to a base
acid dissociation constant (K_a)	the equilibrium constant for the dissociation of a weak acid in water; for the weak acid HA, $K_a = \dfrac{[H^+(aq)][A^-(aq)]}{[HA(aq)]}$
acidic buffer	one that maintains pH at a value below 7
acidity constant	see *acid dissociation constant*
acidity reaction	a reaction of a metal aqua-ion in which an O—H bond in a co-ordinated water molecule is broken, releasing H^+; also known as the *hydrolysis reaction*
activation energy	the minimum energy required for a reaction to occur
adsorption	the process by which a substance (a liquid or a gas) is weakly bonded to and held in place on a solid surface
affinity	see *electron affinity*
amphoteric character	the ability to react with both acids and bases
amphoteric oxide	an oxide that is capable of reacting with both acids and bases.
amphoterism	the property of being able to react with both acids and bases
aqua ion	a metal ion surrounded by water ligands
autocatalysis	a process in which a reaction is catalysed by one of its products
basic buffer	one that maintains pH at a value above 7
Beer–Lambert law	$A = \varepsilon cl$ where absorbance (A), concentration (c), cell path-length (l) and molar absorption coefficient (ε) are linked
bidentate	able to donate a pair of electrons from each of two separate atoms
bond dissociation enthalpy ($\Delta_{diss}H^{\ominus}$)	the standard molar enthalpy change for the breaking of a covalent bond in a gaseous molecule to form two gaseous free radicals, e.g. for the process $Cl_2(g) \rightarrow 2Cl\cdot(g)$ *or* $CH_4(g) \rightarrow \cdot CH_3(g) + H\cdot(g)$
Born–Haber cycle	a Hess cycle/energy level diagram in which the enthalpy of formation of an ionic solid is broken up into a number of steps; it is commonly used for the calculation of lattice enthalpy
Brønsted–Lowry acid	a proton donor
Brønsted–Lowry base	a proton acceptor
buffer range	the pH range over which a weak acid/base can show buffer action
buffer region	the concentration range over which a weak acid/base can show buffer action
buffer solution	one that resists changes in pH on addition of small amounts of acid or base, or on dilution
carboxyhaemoglobin	a very stable complex of haemoglobin; the affinity of haemoglobin for carbon monoxide is much greater than its affinity for oxygen, so that this complex is formed in preference to *oxyhaemoglobin*, thereby inhibiting the uptake of oxygen by red blood cells and making carbon monoxide a very dangerous gas
catalyst	a substance that alters the rate of a reaction without itself being consumed
catalyst poison	an unwanted contaminant or waste product which is adsorbed too strongly on to a catalyst surface, thereby preventing it from acting efficiently
catalytic converters	used in cars; contain catalysts capable of converting harmful gaseous products into less harmful ones
cell convention	the cell diagram is written with the more positive electrode (the one at which reduction occurs) shown as the right-hand electrode

cell diagram	has two electrodes back to back, joined by a salt bridge that is conventionally denoted by two vertical bars, e.g. for a zinc/copper cell, the diagram is $Zn(s) \mid Zn^{2+}(aq) \parallel Cu^{2+}(aq) \mid Cu(s)$
cell EMF	is given by *cell EMF* $= E^{\ominus}(R) - E^{\ominus}(L)$
cell potential	see *standard EMF*
cell reaction	with the more positive electrode shown as the right-hand electrode, the cell reaction goes in the forward direction
charge-to-size ratio	the ratio of the overall charge on an ion to its size
chelate effect	the effect of driving a reaction towards products in the reaction when a bidentate or a multidentate ligand reacts with a co-ordination compound surrounded by unidentate ligands (see *entropy-driven reaction*)
cisplatin	an anti-cancer drug with the formula $[Pt(NH_3)_2Cl_2]$
colorimeter	a simple spectrophotometer used to measure the intensity of colours
combined heat and power (CHP) systems	domestic electricity generation systems which use up waste heat to provide local heating; these systems are made more cost-effective by feeding excess electrical power, during 'idle' periods, into the electricity grid, thus running meters 'backwards'
complex	see *co-ordination compound*
Contact process	the industrial process for converting a mixture of sulfur dioxide and oxygen into sulfur trioxide; uses a vanadium(V) oxide catalyst
co-ordinate bond	a covalent bond formed when the pair of electrons originate from one atom; a straight arrow is drawn to show the origin of the electron pair in the co-ordinate bond, e.g. as in $F_3B \leftarrow NH_3$
co-ordination compound	a compound formed when one or more ligands bond to a metal ion, commonly called a *complex*
co-ordination number	the number of atoms bonded to a metal ion
couple	see *redox couple*
d-block elements	elements in the Periodic Table that have their highest energy electrons in a d sub-level
diprotic acid	one that forms two moles of protons per mole of acid, e.g. H_2SO_4
direction of a cell reaction	a cell reaction goes forwards, in the direction written, if the corresponding *cell EMF* is positive
disposable cell	see *non-rechargeable cell*
disproportionation	the term applied to a reaction in which the same species is simultaneously oxidised and reduced
$EDTA^{4-}$	the bis[di(carboxymethyl)amino]ethane ion, a *hexadentate* ligand with many uses in analytical and industrial chemistry; forms very stable complexes
electrochemical cell	contains two electrodes immersed in an electrolyte
electrochemical series	a list of standard electrode potentials, $E^{\ominus}$, arranged in order of their numerical values
electrode compartment	an electrode immersed in its associated electrolyte
electrode reaction	the half-reaction, oxidation or reduction, that occurs in a given cell compartment, e.g. for the process $Zn(s) \rightarrow Zn^{2+}(aq) + 2e^-$ *or* $Cu^{2+}(aq) + 2e^- \rightarrow Cu(s)$
electrode representation	the notation used to represent an electrode in a cell diagram, e.g. the standard Cu/Cu^{2+} couple is written: $Cu(s) \mid Cu^{2+}(aq, 1.00\ mol\ dm^{-3})$ $E^{\ominus} = +0.34$ V at 298 K
electron affinity ($\Delta_{ea}H^{\ominus}$)	the standard molar enthalpy change for the addition of an electron to an isolated atom in the gas phase, e.g. for the process $Cl(g) + e^-(g) \rightarrow Cl^-(g)$
endothermic change	one in which heat energy is taken in
end-point	the point during a titration when the colour of an indicator lies half-way between the acid and base colours, i.e. $[HIn] = [In^-]$ for the indicator HIn
enthalpy change (ΔH)	the amount of heat released or absorbed when a chemical or physical change occurs at constant pressure (see also *standard enthalpy change*)

enthalpy of atomisation $(\Delta_{at}H^\ominus)$	the standard enthalpy change for the formation of one mole of gaseous atoms from an atomic element in its standard state, e.g. for the process $Na(s) \rightarrow Na(g)$
enthalpy of formation	see *standard enthalpy of formation*
enthalpy of fusion $(\Delta_{fus}H^\ominus)$	the standard molar enthalpy change when a solid forms a liquid at its melting point, e.g. for the process $H_2O(s) \rightarrow H_2O(l)$
enthalpy of hydration $(\Delta_{hyd}H^\ominus)$	the standard molar enthalpy change for the formation of aqueous ions from gaseous ions, e.g. for the process $Mg^{2+}(g) \xrightarrow{water} Mg^{2+}(aq)$
enthalpy of lattice dissociation $(\Delta_L H^\ominus)$	the standard molar enthalpy change for the separation of a solid ionic lattice into its gaseous ions, e.g. for the process $NaCl(s) \rightarrow Na^+(g) + Cl^-(g)$, which is *endothermic*
enthalpy of lattice formation $(\Delta_L H^\ominus)$	the standard molar enthalpy change for the formation of a solid ionic lattice from its gaseous ions, e.g. for the process $Na^+(g) + Cl^-(g) \rightarrow NaCl(s)$, which is *exothermic*
enthalpy of solution $(\Delta_{sol}H^\ominus)$	the standard molar enthalpy change that occurs when an ionic solid dissolves in enough water to ensure that the dissolved ions are well separated and do not interact with one another, e.g. for the process $NaCl(s) \xrightarrow{water} Na^+(aq) + Cl^-(aq)$
enthalpy of sublimation $(\Delta_{sub}H^\ominus)$	the standard molar enthalpy change that occurs on sublimation, when a solid changes directly to a gas without forming a liquid phase
enthalpy of vaporisation $(\Delta_{vap}H^\ominus)$	the standard molar enthalpy change when a liquid forms a gas at its boiling point, e.g. for the process $H_2O(l) \rightarrow H_2O(g)$
entropy	a measure of the disorder in a system
entropy-driven reaction	one where a large positive entropy change dominates a much smaller enthalpy change, making the value of ΔG for the reaction large and negative, resulting in a very favourable (feasible) process, e.g. reactions of $EDTA^{4-}$
equilibrium constant (K_c)	the ratio of concentrations of products and reactants raised to the powers of their stoichiometric coefficients; e.g. for the reaction $3A = 2B + C$ $\quad K_c = \dfrac{[B]^2[C]}{[A]^3}$
equilibrium constant (K_p)	the ratio of partial pressure of gaseous products and reactants raised to their stoichiometric coefficients; e.g. for the equilibrium $2A + B \rightleftharpoons 3C$ $\quad K_p = \dfrac{p(C)^3}{p(A)^2 p(B)}$
equivalence point	the point on a titration curve at which stoichiometrically equivalent amounts of acid and base have been mixed together
excited state	an energy state higher than the ground state
exothermic change	one in which heat energy is given out
***E–Z* stereoisomerism**	also known as geometrical or *cis–trans* isomerism
***E–Z* stereoisomers**	arise due to restricted rotation about a carbon–carbon double bond when the two pairs of attached substituents can be arranged in two different ways; this type of stereoisomerism can also arise in appropriately substituted transition-metal complexes
feasibility temperature	the equilibrium temperature, T, at which $\Delta G = 0$: $\quad T = \dfrac{\Delta H^\ominus}{\Delta S^\ominus}$
feasible (spontaneous) change	one that has a natural tendency to occur without being driven by any external influences; in terms of Gibbs free-energy, $\Delta G \leq 0$ for feasible change
first ionisation enthalpy (energy) $(\Delta_i H^\ominus)$	the standard molar enthalpy change for the removal of an electron from an atom in the gas phase to form a positive ion and an electron, both also in the gas phase, e.g. for the process $Na(g) \rightarrow Na^+(g) + e^-(g)$
free radical	a species that results from the homolytic fission of a covalent bond; it contains an unpaired electron
fuel cell	one which produces electrical power from an external supply of a fuel and an oxidant

gas electrode	an inert metal (usually platinum) surrounded by a gas in equilibrium with a solution of its ions (see *standard hydrogen electrode*)
Gibbs free-energy change ($\Delta G^\ominus$)	determines the direction of spontaneous change by combining the influences of enthalpy and of entropy, through the relationship $\Delta G^\ominus = \Delta H^\ominus - T\Delta S^\ominus$
Gibbs free-energy change at equilibrium	for a system at equilibrium, $\Delta G = 0$
ground state	the lowest energy state of electrons in a species, e.g. of d electrons in a transition-metal complex
Haber process	the industrial process for converting a mixture of nitrogen and hydrogen into ammonia; uses an iron catalyst
haemoglobin	an octahedrally co-ordinated iron(II) complex, responsible for the red colour of blood and for the transport of oxygen by red blood cells from one part of the body to another
half-equation	a balanced equation for an oxidation or a reduction that shows a species gaining or losing electrons, e.g. for the process $MnO_4^- + 8H^+ + 5e^- \longrightarrow Mn^{2+} + 4H_2O$
half-equivalence	when exactly one-half of the equivalence volume of a base or acid has been added to an acid or base.
heterogeneous catalysis	proceeds through the adsorption of reactants on to a catalytic surface
heterogeneous catalyst	acts in a different phase from the reactants
heterogeneous system	one with the species present in different phases
heterolytic fission	formation of ions when a covalent bond breaks with an unequal splitting of the bonding pair of electrons, e.g. for the process $HCl \longrightarrow H^+ + Cl^-$
hexadentate ligand	having six electron pairs capable of forming co-ordinate bonds (see *EDTA^{4-}*)
homogeneous catalysis	proceeds through the formation of an intermediate species in the same phase as the reactants
homogeneous catalyst	acts in the same phase as the reactants
homogeneous system	one with all species present in the same phase
hydrogen electrode	corresponds to the redox couple H^+/H_2 at a platinum surface
hydrolysis reaction	see *acidity reaction*
indicator	usually a weak organic acid with strongly coloured acid (HIn) and base (In$^-$) forms
inert support medium	an unreactive solid used to dilute a reagent or a catalyst
ionic product of water (K_w)	$K_w = [H^+(aq)][OH^-(aq)]$
irreversible cell	see *non-rechargeable cell*
isomers	molecules with the same chemical formula but in which the atoms are arranged differently (see *structural isomerism* and *stereoisomerism*)
IUPAC convention for electrode reaction half-equations	all redox half-equations are written as *reductions* (electron gain)
lattice enthalpy	see *enthalpy of lattice dissociation* and *enthalpy of lattice formation*
Le Chatelier's principle	states that a system at equilibrium will respond to oppose any change imposed upon it
ligand	an atom, ion or molecule capable of donating one or more pairs of electrons to a metal ion; a ligand acts also as a Lewis base or a nucleophile
ligand substitution	replacement of one or more ligands by others
linear complex	a metal ion with two ligands in a straight line on either side; the two bond angles are each $180°$
metal electrode	conventionally written as $M^{n+}(aq)/M(s)$, or the reverse
mixed catalysts	those prepared by mixing two or more catalytic substances
mole fraction	the mole fraction x of a component A in a gas mixture is defined as: $$x_A = \frac{\text{number of moles of } A}{\text{total number of moles of all compounds}}$$

monoacidic base	one that forms one mole of hydroxide ions per mole of base, e.g. NaOH
monodentate	able to donate a pair of electrons from one atom only
monoprotic acid	one that forms one mole of protons per mole of acid, e.g. HCl
multidentate ligand	able to donate a pair of electrons from each of several separate atoms
non-rechargeable cell	one not intended to be recharged by an electric current; also known as a *disposable cell*
nucleophile	an electron-rich molecule or ion able to donate a pair of electrons
octahedral complex	a central metal ion with six ligands lying at the corners of a octahedron; the six bond angles in a regular octahedron are each 90°
optical isomers	stereoisomers (enantiomers) which rotate the plane of plane-polarised light equally but in opposite directions
overall equation for a redox reaction	obtained by adding together two half-equations and balancing the numbers of electrons gained and lost in each half-equation
oxidation	the process of electron loss
oxidation state	the charge a central atom in a complex would have if it existed as a solitary simple ion without bonds to other species
perfect ionic model	ionic crystals consisting of oppositely-charged discrete spherical ions that interact exclusively through electrostatic attractions and repulsions
Period 3 elements	elements in the third period (row) of the Periodic Table
pH	logarithmic expression of the proton concentration in aqueous solution $pH = -\log_{10}[H^+(aq)]$
pH at half-equivalence	at half-equivalence, $pH = pK_a$ for a weak acid
pK_a	logarithmic expression of the acid dissociation constant in aqueous solution $pK_a = -\log_{10}K_a$
Planck's constant (h)	the constant of proportionality between the energy (ΔE) of absorbed light and its frequency (v) in the relationship $\Delta E = hv$, where h has the value 6.63×10^{-34} J s
polarising power	the ability of a metal ion to distort the electron cloud of a neighbouring anion
proton acceptor	a substance that accepts protons in a chemical reaction
rechargeable cell	one designed to be recharged by an electric current
redox couple	a shorthand way of writing a reduction half-equation, e.g. for the process $Cu^{2+}(aq) + 2e^- \rightarrow Cu(s)$, the redox couple is Cu^{2+}/Cu
redox electrode	one at which two oxidation states of a given element undergo a reduction reaction at an inert metal surface, e.g. for the Fe^{3+}/Fe^{2+} couple the electrode is written $Pt(s) \mid Fe^{2+}(aq), Fe^{3+}(aq)$
redox potential	the standard electrode potential for a reduction process, e.g. $Cu^{2+}(aq) + 2e^- \rightarrow Cu(s)$
redox titration	used in volumetric analysis to determine the concentration of either an oxidising agent or a reducing agent, e.g. iron(II) with manganate(VII) ions
reduction	the process of electron gain
reference temperature	under standard conditions, this temperature is most commonly taken to be 298 K
salt bridge	an electrolyte solution used to complete electrical contact between two electrode compartments; it allows the transfer of ions between compartments
second ionisation enthalpy	the standard molar enthalpy change for the removal of an electron from a singly positively charged ion in the gas phase to form a gaseous di-positive ion and an electron, both also in the gas phase, e.g. for the process $Na^+(g) \rightarrow Na^{2+}(g) + e^-(g)$
sequestering ability	(*to sequester* – literally *to give up for safekeeping*) – the property of a substance to keep metal ions in solution even when anions which normally cause precipitation (such as OH⁻ or CO_3^{2-}) are added
spontaneous (feasible) change	one that has a natural tendency to occur without being driven by any external influences; in terms of Gibbs free-energy, $\Delta G \leq 0$ for spontaneous change

square-planar complex	a central metal ion with four ligands lying in one plane at the corners of a square; the four bond angles in such a square are each 90°
standard amount	the mole
standard changes	changes measured under standard conditions and signified by the use of a superscript *plimsoll sign,* $\ominus$
standard conditions	refer to a standard pressure of 100 kPa at a stated reference temperature (most commonly 298 K); for solutions, also at a concentration of 1.00 mol dm^{-3}
standard electrode potential ($E^\ominus$)	the standard potential of a cell measured with the standard hydrogen electrode as the left-hand electrode and the unknown as the right-hand one
standard EMF	the potential difference between the electrodes of a standard electrochemical cell measured under zero-current conditions
standard enthalpy change ($\Delta H^\ominus$)	the change in enthalpy when reactants in their standard states form products also in their standard states
standard enthalpy of formation ($\Delta_f H^\ominus$)	the enthalpy change under standard conditions when one mole of a compound is formed from its elements with all reactants and products in their standard states
standard entropy ($S^\ominus$)	an absolute measure of entropy, based on zero entropy at zero Kelvin
standard hydrogen electrode	has a defined potential of 0 V and an electrode representation: $Pt(s) \mid H_2(g, 1 \text{ bar}) \mid H^+(aq, 1.00 \text{ mol dm}^{-3})$ $E = 0$ at 298 K
standard state	the pure, most stable, form of a substance at a given temperature and 100 kPa
stereoisomerism	occurs when molecules with the same structural formula have bonds arranged differently in space (see *E–Z stereoisomerism* and *optical isomerism*)
stereoisomers	are compounds which have the same structural formula but have bonds arranged differently in space
stoichiometric point	see *equivalence point*
strong acid/base	one that is (almost) completely dissociated in aqueous solution
sublimation	the process by which a substance changes directly from a solid to a gas without forming a liquid phase, e.g. for the process $I_2(s) \longrightarrow I_2(g)$
substitution reaction	one in which one or more ions or molecules are replaced by other ions or molecules, e.g. the replacement of one ligand by another in a reaction of a transition-metal complex
surface-to-mass ratio	the ratio of the area of a surface to the mass of the substance covering it
tetrahedral complex	a central ion with four ligands lying at the corners of a tetrahedron; the four bond angles in a regular tetrahedron are each 109.5°
titration curve	a plot of the pH of an acid/base against the volume of base/acid added
titrations using potassium manganate(VII)	when used as an oxidising agent in acidic solution, no additional indicator is required; at the end-point, the presence of even a slight excess of the dark purple manganate(VII) ion is used to indicate complete reaction as the solution turns pink
Tollens' reagent	contains the complex ion $[Ag(NH_3)_2]^+$ which, with aldehydes (but not ketones), is reduced to silver on warming
transition element	an element having an incomplete d (or f) sub-level either in the element or in one of its common ions
transition-metal complex redox reaction	one in which a transition-metal ion changes its oxidation state, either being oxidised or being reduced
visible spectrophotometer	a device that uses visible light of varying frequencies to measure the amount of light absorbed by a coloured solution; the absorbed amount is proportional to the concentration of the absorbing species in the solution under test
weak acid approximation	when K_a is small, $K_a \approx \dfrac{[H^+]^2}{[HA]_{\text{tot}}}$
weak acid/base	one that is only partially dissociated in aqueous solution

Index